*L I F E L I N E S*

# HUMAN MICROBIOLOGY

D0303571

■ Simon P. Hardy

School of Pharmacy and Biomolecular Science,
University of Brighton, UK

London and New York

First published 2002 by Taylor & Francis
11 New Fetter Lane, London EC4P 4EE

Simultaneously published in the USA and Canada
by Taylor & Francis Inc,
29 West 35th Street, New York, NY 10001

*Taylor & Francis is an imprint of the Taylor & Francis Group*

© 2002 Simon P. Hardy

Typeset in Perpetua by Wearset Ltd, Boldon, Tyne and Wear
Printed and bound in Great Britain by TJ International Ltd, Padstow, Cornwall

*British Library Cataloguing in Publication Data*
A catalogue record for this book is available from the British Library

*Library of Congress Cataloging in Publication Data*
A catalog record for this book has been requested

ISBN 0-415-24199-5 (hbk)
ISBN 0-415-24168-5 (pbk)

# CONTENTS

# SERIES EDITOR'S PREFACE

Teaching programmes in universities now are generally arranged in collections of discrete units. These go under various names such as units, modules, or courses. They usually stand alone as regards teaching and assessment but, as a set, comprise a programme of study. Usually around half of the units taken by undergraduates are compulsory and effectively define a 'core' curriculum for the final degree. The arrangement of teaching in this way has the advantage of flexibility. The range of options over and above the core curriculum allows the student to choose the best programme for her or his future.

The Lifeline series provides a selection of texts that can be used at the undergraduate level for subjects optional to the main programme of study. Each volume aims to cover the material at a depth suitable to a particular level or year of study, with an amount of material appropriate to around one-quarter of the undergraduate year. The concentration of life science subjects in the Lifeline series reflects the fact that it is here that individual topics proliferate.

Suggestions for new subjects and comments on the present volumes in the series are always welcomed and should be addressed to the series editor.

*John Wrigglesworth*
*London, March 2000*

# INTRODUCTION

Recent catastrophic earthquakes have resulted in television images of disinfectant being sprayed over the rubble of collapsed houses whilst earnest reporters tell us of these attempts to prevent outbreaks of typhoid and cholera. Not an encouraging picture if you realise that those infections are not caught by inhalation from dry and dusty rubble.

Every person continuously interacts with micro-organisms throughout their lifetime. The extent to which this encounter will be a pleasant or harrowing experience will vary enormously. Certainly all of us will experience unpleasant infections of some sort, if only the common cold. More sobering is the fact that infectious diseases still account for the greatest number of deaths in the world, despite being preventable. Irrespective of the tremendous advances in our understanding of microbial physiology, infectious disease will have the final say for most of mankind. Pneumonia will not only kill millions of children in the Developing World this year but dispatch a considerable fraction of the elderly population in the West as well.

There may be several reasons why organisms that individually are invisible to the naked eye have captivated the attention of many people. Perhaps it is the disproportionate impact of the organism compared with its size? An interest in infectious diseases is often what draws attention to the activity of bacteria, but once hooked, the physiology of microscopic life then provides lifelong fascination. It is hoped that this little book will stimulate the same sequence of events. The aim is to introduce micro-organisms to someone who has been studying biology but as yet has not delved into the fascinating world of microscopic organisms and leave them appreciating why our struggles with infections do not seem to be getting any easier. The book attempts to deal with mathematics such that nervous readers with an uncertain grasp of the topic can keep reading. Although we are warned that every formula loses a significant number of readers, the mathematics presented is largely simple rearrangements of formulae that should fox few people (given a pencil and five minutes' thought).

*Human Microbiology* attempts to distil the key principles of medical microbiology for undergraduate students of biomedical science and biology. The condensation of all things infectious into 250 pages will also be useful for students of human and veterinary medicine as well as other professions allied to medicine. The book seeks to provide a basis for understanding the biology of micro-organisms that interact with humans but does not seek to go through long (and increasing) lists of organisms and diseases as found in many huge, exhaustive textbooks of microbiology. As wonderful as they are, they are for the converted and are not entirely appropriate for modular degree programmes. Hence, bacterial genetics and biochemistry have simply been outlined and more suitable modular texts should be investigated. The book is divided into two sections. The first describes the basic physiology of the three important groups of micro-organisms – bacteria, viruses and fungi – and the second section examines the underlying themes that describe their interactions with humans. Microbiology is more than a description of micro-organisms, and chapters on microbial taxonomy and methods of controlling microbes have been included. Far from the image of a tired, outdated discipline, there is renewed interest in taxonomy. As a product of the newer methods in analysing gene sequences the old views on the origins of all living things (no less!) have been turned on their head. The chapter on microbial control measures looks at how microbial diseases have been dealt with and, although earthquakes have not been covered specifically, design strategies to control microbial infection are considered.

Thanks need to be placed in print as a permanent reminder that the book was only possible with the unlimited and unflagging support of Sharon and Miles. Likewise, the generous instruction and effort of my tutors and colleagues, David Jarvis, Mark Holland, John Foord, Marcus Allen and Paul Everest deserves recognition, if not medals. Appreciative thanks go out to Joyce Storey at the Aldrich Library, the University of Brighton for tracking down articles and books and the encouraging discussions with Dilys Alam at Taylor & Francis. As is customary, the errors, distortions and omissions are entirely my own. With scientific writing, there is always a tension between fine detail and unsubtle generalisation. One only hopes that the text serves to stimulate further enquiry in the reader.

In the 1970s, the suggestion was widespread that infectious diseases as a speciality was going to become a minority subject in medical schools, perhaps of importance only in developing countries. Pharmaceutical companies down-sized their antibiotic R&D programmes in favour of other illnesses. Easy to say now, but this has proved either woefully naïve, optimistic or stupid. This book has been written against a backdrop of bioterrorism in the USA as anthrax spores sent through the post have caused several deaths and epidemic foot and mouth disease in England that has devastated sheep, cattle and pig numbers in the country, sending numerous farms into bankruptcy. Clearly, viral diseases and epidemiology need to be included in university economics courses if such disasters are to be adequately controlled. How does a port authority inspector spot morbilliviruses in lorries? If, when we get the call to rescue the nation from untamed viruses, the population is wiped out, we should at least be able to console ourselves that we knew how and why it happened.

# I ■ INTRODUCTION TO MICRO-ORGANISMS

Bacteria must have got something right. If the earth was formed approximately 4 to 5 billion ($1 \times 10^9$ million) years ago, it is estimated that bacteria appeared a billion years later (Figure 1.1), whereas, in relative terms, animals and plants have only just appeared. The diversity of bacteria is difficult to comprehend fully since it is estimated that around 90 per cent of all bacterial types have not yet been successfully grown in the laboratory. Microbiologists who concentrate on bacteria that interact with humans will encounter a tiny, specialist collection of organisms that is not representative of the tremendous variation that can be seen in more diverse environments. Bacteria have a variety of shapes, which presumably reflects the extreme range of habitats they occupy. Bacteria thrive at temperatures that range between below freezing and greater than the boiling point of water. Bacteria grow faster than all other organisms, and utilise a broad spectrum of chemicals as nutrient sources and energy sources. Yet, at the other extreme, under adverse conditions (for example, low available water and nutrient levels) bacteria have survival strategies of which the production of **endospores** yields the most resilient form.

Endospores ('spores') are produced by both bacteria and fungi. They are thick-walled forms of the organism that are resistant to adverse conditions such as desiccation and heating. Spores can develop into new organisms without fertilisation, when conditions improve.

As microscopic organisms, bacteria can reach and occupy all habitats that support

• **Figure 1.1** History of the planet Earth. Various key events in the development of the earth are shown. Microbes have been present for over 3 billion years compared with mammals, comparative juveniles of less than a million years old

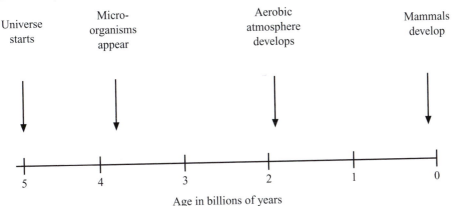

| Universe starts | Micro-organisms appear | Aerobic atmosphere develops | Mammals develop |

|---|---|---|---|---|---|
| 5 | 4 | 3 | 2 | 1 | 0 |

Age in billions of years

life. The dispersal of bacteria can be readily achieved in fluids either passively or actively via mechanisms for motility (such as flagella), but bacteria can also exploit airborne transmission. Bacteria may be microscopic in size but their influence is disproportionately widespread. Rather than being seen as single cells battling it out against other competing organisms, bacteria form communities (mixed populations) and communicate with each other so as to best exploit the resources that exist between them. The genetic information between bacterial communities can be transferred between organisms. Complex interactions occur as different bacteria utilise the nutrients between themselves such that little is not utilised.

Later we will examine some of the mechanisms used to exploit eukaryotes as habitats. It is interesting to see which of the key features outlined here play a critical role in their ability to infect humans:

- rapid growth rate,
- metabolic versatility,
- cell size,
- resistance to adverse conditions.

Having indicated the diversity of bacteria in general, individual species can be considered to adopt **generalist** or **specialist** survival strategies. Generalists occupy a diverse range of habitats whereas specialists will be found only in particular sites. In stable environments it is expected that specialists will give less attention to the production of resistant, dormant stages such as endospores if conditions are being kept relatively constant within a mammalian host. Sporulation (the production of spores) will cost in terms of genes and energy, and illustrates a tension between generalists wishing to accumulate functional genes and specialists tending to reduce genome size in order to become as efficient as possible.

## ■ 1.1 BACTERIAL SIZE

Individual bacterial cells are invisible to the naked eye. Only when their numbers reach $10 \times 10^6$ (10 million) do they become visible and appear as a cloudiness (turbidity) when in suspension, or as individual colonies up to a couple of millimetres in diameter on an agar plate or on the surface of old yoghurt in the back of the fridge.

Small things, quite obviously, do not take up much space; hence, large numbers of micro-organisms can be found in small volumes. Of greater significance to the study of infectious diseases is the fact that they are invisible to the naked eye. Only with an understanding of the characteristics of micro-organisms can effective measures be taken to reduce the transmission of the associated diseases.

Length is referenced to the metre (m). One micron ($\mu$m) is a thousand fold smaller than a millimetre ($1 \times 10^{-3}$ mm) or $1 \times 10^{-6}$ of a metre ($1 \times 10^{-6}$ m).

In general, the sizes of bacteria isolated from humans fall within a limited range. Spherical bacteria (cocci) are between 0.5–1.5 $\mu$m in diameter. Rod shaped or cylindrical bacteria (bacilli or 'rods') have diameters of 0.2–1 $\mu$m by 0.5–5 $\mu$m in length. With the resolution of the human eye limited to 0.1 mm, single bacterial cells have to be viewed using light microscopy with a magnification of ×400. Obviously, electron microscopy will yield greater detail. A scanning electron microscope can magnify up to 100,000 fold.

## ■ 1.1.1 THE 'TYPICAL' BACTERIUM

The concept of a typical bacterium is somewhat perverse since the variation in bacterial shapes that exist within nature is dizzying. Nonetheless, those that infect humans are

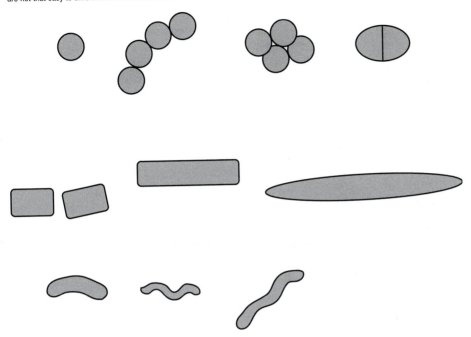

• **Figure 1.2** Diagrammatic representation of some bacterial shapes. The top row shows various arrangements of cocci. The middle row shows the variation in rod-shaped bacteria and the bottom row shows the curved and spiral-shaped bacteria. Some of the shapes are not that easy to differentiate between cocci and rods

relatively restricted in shape and function. Of those that infect humans there are essentially three forms of bacterial shape: spheres, cylinders and spirals, as shown in Figure 1.2. Spheres are called 'cocci', cylinders are called 'rods' or 'bacilli'. There are variations on all three shapes, for example very short rods that can be difficult to differentiate between cocci and rods are called 'cocco-bacilli', resembling coffee beans in shape. An example of the variation seen in bacterial patterns observed under the microscope is that of the genera Staphylococcus and Streptococcus. Collections of staphylococci tend to form clusters of cells (the name is derived from the Greek, meaning clusters of grapes or berries) whereas streptococci tend to form chains. Figure 1.3 indicates how the plane of division in the dividing bacterium will influence the pattern formed by the bacteria.

### ■ 1.1.2 DIFFUSION OF NUTRIENTS AND GASES

The variation in size of micro-organisms is not unlimited. The efficiency with which the organism can accumulate nutrients and dispose of waste material through the cytoplasmic membrane will restrict expansion. Many key metabolites (e.g. oxygen) pass passively across the cell wall and cell membrane into the cytosol. The surface area to volume ratio (SA/V) is the limiting factor for the extent to which passively diffusing molecules penetrate the cytosol. For spherical organisms the diameter is constrained by the penetration of oxygen meeting the consumption by the organism. Figure 1.4 shows the relationship between cell diameter of a spherical bacterium and surface area to volume ratios and shows that constraints on maximum size are reflected in the rapidly diminishing SA/V. Surface area is a function of the square of the cell radius but the volume increases by a power of 3. Because the SA/V ratio increases as the radius gets

• **Figure 1.3** Plane of division in cocci. (A) Streptococci tend to form chains of cocci, reflecting the repeated plane of division through the vertical axis. Whereas, in staphylococci (B), the plane of division changes and the cells divide in different planes yielding tetrads and clusters

(A)

(B)

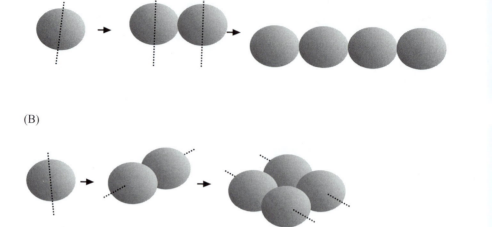

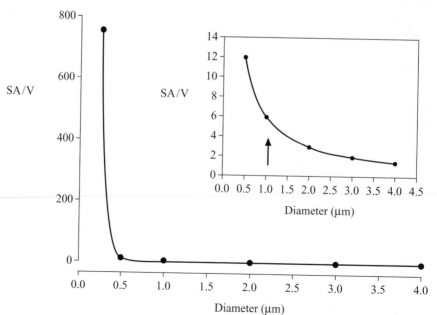

• **Figure 1.4** Surface area to volume ratio for a sphere with diameter between 0.25 to 4 μm diameter. The main graph shows the dramatic fall in SA/V ratio when the diameter is greater than 0.25 μm. The inset shows in more detail the fall in SA/V between 0.5 to 4 μm diameter spheres. Note that *Staphylococcus aureus*, with a diameter of 1 μm (arrow), appears to fall in the middle, suggestive of a trade off between being too big to permit adequate diffusion and too small to accommodate all the necessary cell components smaller, the SA has greater effect on smaller organisms. Any size greater than 1 μm in diameter reduces the SA/V to less than 5 to 1, which presumably impedes the functioning of the bacterial cell. Figure 1.5 shows that the SA/V ratio for bacilli is unaffected by increases in length, if the diameter remains unchanged.

The physical packing of the nucleic acid (see below, p. 15) and cytoplasmic components such as polysomes into a bacterial cell will also limit the minimum size achievable. Larger single-celled eukaryotic organisms such as protozoa overcome the insufficient cytoplasmic membrane through the use of specialised organelles (e.g. endoplasmic reticulum, mitochondria). With increasing size comes the need to use specialised systems to transport materials throughout the organism because the SA/V ratio is inadequate.

## ■ 1.2 BACTERIAL STRUCTURES

Figure 1.6 represents a bacterial cell in which the most important features we will discuss are shown.

## ■ 1.2.1 CELL WALL

Whilst most higher organisms support themselves with a skeleton onto which tissues are placed, micro-organisms place the protective skeleton on the outside as a rigid cell

• **Figure 1.5** Surface area to volume ratio (SA:Vol) in cocci and bacilli. Note how the increasing diameter of a sphere (coccus) reduces the SA:Vol ratio whereas in bacilli the increasing length has no such effect

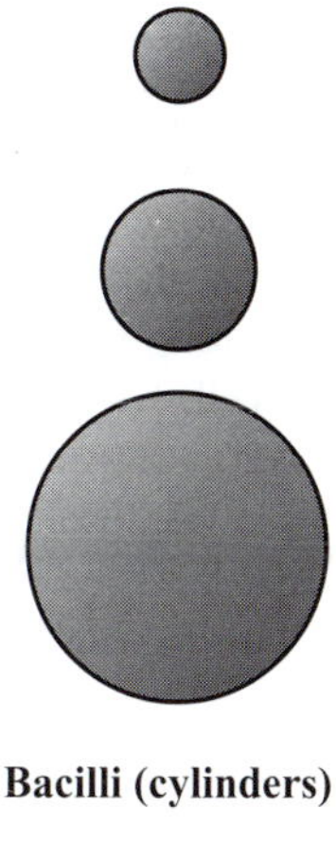

| Cocci (spheres) | Diameter (μm) | | SA | Vol | SA:Vol |
|---|---|---|---|---|---|
| | 1 | | 3.14 | 0.5 | 6 |
| | 2 | | 12.6 | 4.2 | 3 |
| | 4 | | 50.2 | 33.5 | 1.5 |

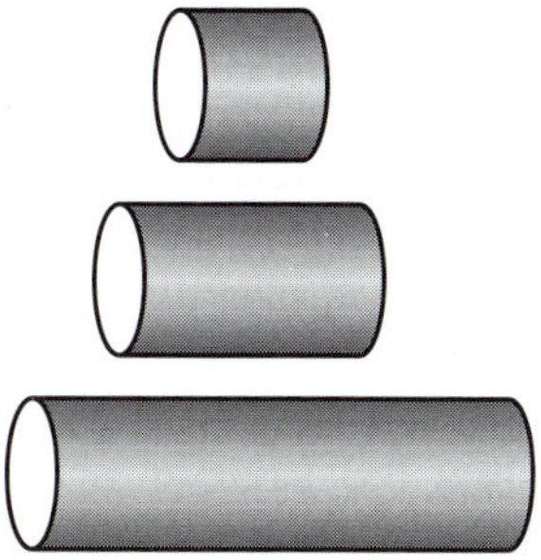

| Bacilli (cylinders) | Diameter (μm) | Length (μm) | SA | Vol | SA:Vol |
|---|---|---|---|---|---|
| | 2 | 2 | 12.6 | 6.3 | 2 |
| | 2 | 4 | 25.12 | 12.56 | 2 |
| | 2 | 6 | 37.7 | 18.8 | 2 |

• **Figure 1.9** Gram negative cell wall. The presence of the outer membrane has created an extra hydrophobic membrane outside of the peptidoglycan layer. To allow access to the cytoplasmic membrane, water-filled pores (porins) are necessary. Note the thin layer of peptidoglycan compared with Gram positive cell walls

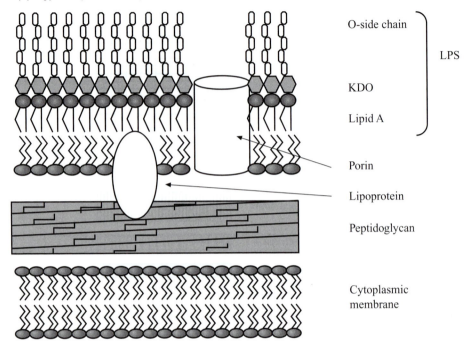

O-side chain

LPS

KDO

Lipid A

Porin

Lipoprotein

Peptidoglycan

Cytoplasmic membrane

glycopeptides that is constructed in a helix to form a cylindrical scaffold-like structure in bacilli or as a sphere in cocci. The structure of the peptidoglycan confers strength as well as sufficient elasticity to cope with stresses on the wall.

Peptidoglycan is composed of alternating **N-acetylglucosamine (NAG)** and **N-acetylmuramic acid (NAM)** molecules. Both are carbohydrates (glucose) with amino groups attached. The NAG and NAM units are cross-linked through covalent bonding of four unusual amino acids to form a (tetra-) peptide chain (Figure 1.10). Note that NAG and NAM are connected by three covalent bonds, two between the carbohydrates themselves and the third via the peptide side chain. This will confer considerable strength. Chains of carbohydrates are called glucans and the linear disaccharide chains of NAG–NAM resemble the glucans such as chitin found in insects and fungal cell walls (Chapter 4) and cellulose found in plant cell walls. Plant cellulose is also a disaccharide chain but has no covalent crosslinking (the pentapeptide side chains are absent) hence the strength of cellulose lies along the length of the chain. Sideways pressures will easily buckle the chain as seen when a stem of a plant is bent. In contrast, trying to pull a stem apart lengthwise is very difficult. This weakness is corrected for in peptidoglycan because of the peptide side chains.

The composition of the peptidoglycan is conserved across most Gram negative bacteria in contrast to Gram positive cell walls where the structural components vary in the composition of the interpeptide bridge. The protection to the organism conferred by the peptidoglycan is illustrated by the fact that the peptidoglycan uses D-isomer amino acids in the interpeptide bridge. This protects the cell wall against damage (hydrolysis) by protease enzymes that attack L-isomers (Box 1.2).

Furthermore, whereas NAG is found in insect chitin, NAM is unique to bacteria.

• **Figure 1.10** The peptidoglycan of (A) *Escherichia coli* and (B) *Staphylococcus aureus*. Note that the arrangement differs in the cross-linking arrangement.

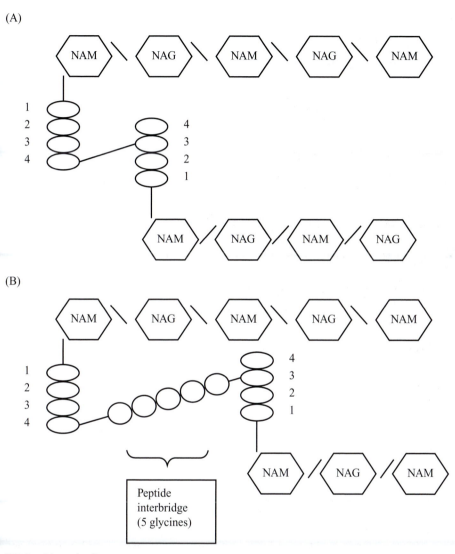

(A)

(B)

NAM: N-acetylmuramic acid
NAG: N-acetylglucosamine
1: L-alanine
2: D-glutamic acid
3: diaminopimelic acid
4: D-alanine

This helps protect the bacterium from hydrolytic degradation by other organisms. Interference with the structure of peptidoglycan may result in cell death, hence it is a suitable target for attack by antibiotics (most notably penicillins). Similarly lysozyme, a natural antibacterial protein secreted in mucous secretions and in tears, hydrolyses the NAM–NAG linkage.

Gram positive bacteria cell walls also contain large quantities (up to 50 per cent) teichoic acids (polymers of sugars and phosphate: ribitol phosphate or glycerol phosphate) and lipoteichoic acid (teichoic acid and lipid) (Figure 1.8). The teichoic acids are found in the peptidoglycan and have a net negative charge that contributes to the net negative surface charge of Gram positive bacteria. The coupling of a lipid component to teichoic acid forms a lipoteichoic acid and this permits the molecule to anchor the lipid end in the cytoplasmic membrane. Lipoteichoic acid can extend through the peptidoglycan layer and project from the surface of the cell wall where, in some bacteria (*Streptococcus pyogenes* for example), it is known to act as an **adhesin**.

All molecules that act to help bind the organism to a surface are called **adhesins**.

The Gram negative cell wall has a different arrangement (Figure 1.9 and see Table 1.1). In addition to a thinner peptidoglycan layer, Gram negative bacteria possess a second phospholipid bilayer called the outer membrane. Whereas the cytoplasmic membrane is a symmetrical phospholipid bilayer, the outer membrane is asymmetrical in construction. The inner facing leaflet resembles the inner leaflet of a normal phospholipid bilayer like the cytoplasmic membrane but the outer layer is unique to Gram negative bacteria. The outer membrane is called **lipopolysaccharide (LPS)** and consists of three component parts, **lipid A, core oligosaccharide** and **O-specific side chains**. The O-specific side chains are long carbohydrate chains that form surface antigens. Antibodies raised against these O-antigens are used to distinguish between strains of Gram negative bacteria (serotyping, see Chapter 6); for example, in *Escherichia coli* there are at least 300 O-antigen types. The hydrophilic O-side chains help protect the organisms from hydrophobic compounds such as bile salts. In mutants that are defective in synthesis of the side chains the colonies appear rough (hence 'R mutants') and are susceptible (i.e. inhibited) by bile salts unlike the normal wild type strains which have a smooth colonial appearance. The tolerance to bile salts is such that it is used in culture media to help select for the growth of enteric Gram negative bacteria. Mutations in synthesis of Lipid A and core oligosaccharide are lethal.

Lipopolysaccharide is amphipathic in order to bridge the hydrophilic external face with the hydrophobic core that forms the second leaflet of the outer membrane. The Lipid A component will be inserted into the phospholipid outer membrane. The core and the O-side chains will then project outwards from the cell. Surrounding themselves with a hydrophobic lipid bilayer, Gram negative bacteria need to transport hydrophilic solutes and ions across the outer membrane via water-filled transmembrane protein

**Table 1.1 Essential differences between Gram positive and Gram negative cell wall components, expressed as relative abundance**

|  | Peptidoglycan | Teichoic/LTA | Lipopolysaccharide |
| --- | --- | --- | --- |
| Gram positive cell wall | +++ | ++ | − |
| Gram negative cell wall | + | − | +++ |

---

### ■ BOX 1.2 ISOMERS

Isomers are molecules that have the same molecular formula but arranged in a different configuration. Optical isomers are mirror images of themselves. You may recall that carbohydrates exist as D and L forms, with the D 'sugars' being the predominant form in nature. Conversely, L amino acids are the predominant form in nature but bacteria are exceptional in their use of D amino acids. Bacteria also possess the enzymes (**racemases**) that can convert between D and L forms of both carbohydrates and amino acids.

pores termed **porins**. They permit the flow of low molecular weight ions and solutes (nutrients such as sugars) passively across the hydrophobic outer membrane. The permeability of the outer membrane is much greater than that of the cytoplasmic membrane, due mostly to the protein porins that represent approximately 50 per cent of the outer membrane components.

Once across the outer membrane the hydrophilic nutrients are placed in the periplasmic space. Uptake into the cytoplasm then needs to occur via specific active transporters located in the cytoplasmic membrane. Compare this two-tier strategy with Gram positive bacteria that possess no outer membrane.

The medical significance of lipopolysaccharide lies in its toxicity. Whilst not as potent as bacterial protein exotoxins (see Chapter 8), lipopolysaccharide is a toxin (specifically called **endotoxin**), able to trigger numerous host responses which cause fever. The toxic component is the Lipid A. Gram negative bacteria shed fragments of their cell wall, including lipopolysaccharide, and this continual release may contribute to their ability to initiate symptoms and disease.

The physical framework that the cell wall provides for bacteria helps prevent the bacterial cell from bursting when placed in water. Osmosis tells that just as dried prunes swell when placed in water, the same forces will try and act with bacteria when placed in hypotonic solutions (Box 1.3). The water will attempt to move into the cytoplasm to equilibrate the osmotic strength and so produce turgor pressure on the inside of the cell wall, which if disrupted by lysozyme will result in the cell bursting. **Turgor pressure** is a botanical term and hydrostatic pressure is essentially equivalent in meaning. Bacteria that have lost their cell wall can only be kept viable if held in isotonic solutions (usually by adding up to 300 mM sucrose). Organisms that have lost their cell wall are called **protoplasts** or if residual fragments of cell wall are present are called **spheroplasts**.

There is one further arrangement of a bacterial cell wall that has importance to medical microbiologists: the cell wall of members of the genus Mycobacterium. The species in this genus of greatest importance in terms of human disease are *Mycobacterium tuberculosis* and *Mycobacterium leprae*, the aetiological agents of tuberculosis and leprosy respectively. Mycobacteria are characterised by their waxy cell wall due to the presence of **mycolic acids** (Figure 1.11). The waxy coat will repel hydrophilic compounds such as the crystal violet and iodine complex in the Gram stain. In order to get the cell wall to stain, it is necessary to drive the stain into the cell wall by gently heating the slide.

Once the dyes have penetrated they are difficult to extract, resisting elution by weak acids. This resistance to extraction by acid is why mycobacteria are called 'acid fast

> Porins are bacterial equivalents to mammalian membrane transport proteins that mediate the passage of ions and solutes across the cytoplasmic membrane.

---

### ■ BOX 1.3 OSMOTIC PRESSURE

Hypotonic solutions have a lower ionic strength than the reference solution (also called 'a lower osmotic pressure'). Osmotic strength is measured in Osmoles/litre. A 300 mM solution of sucrose will have an osmotic strength of 300 mOsmoles, whereas the same strength solution of NaCl will have 600 mOsmoles/litre ($2 \times 300$ mOsmoles as both sodium and chloride ions will contribute equally). Isotonic solutions are those with equivalent osmolarity to that inside the cell. A hypertonic solution is a higher osmotic strength. Mammalian cells are roughly 300 mOsmolar, hence a physiological and isotonic salt solution will be 150 mM NaCl.

---

### ■ BOX 1.4 MYCOPLASMAS

Mycoplasmas are bacteria without any trace of peptidoglycan cell wall. Yet in contrast to being seen as incomplete bacteria these organisms are a large and apparently successful group of organisms (more appropriately grouped in the class Mollicutes). Mycoplasmas infect most animals and some are important causes of human disease (for example, *Mycoplasma pneumoniae* is a common cause of pneumonia in man). The lack of cell wall makes the organisms highly susceptible to lysis by osmotic shock and the action of detergents but an obligate association with mucous membranes means that they will not encounter such challenges sufficiently regularly. It follows that mycoplasmas are resistant to antibiotics that target the cell wall, typically penicillins. One particular interest in mycoplasmas lies in their small genome (approximately a fifth the number of genes compared with *Escherichia coli*). The obligate parasitic lifestyle has meant that many genes have been discarded (no respiratory chain and electron transport chains) and the organisms are viewed as the best representative of the smallest free-living, self-replicating cell.

---

bacilli'. The classic staining method for mycobacteria using this principle is the **Ziehl–Neilsen** stain (Box 1.5). The hydrophobic mycobacterial cell wall confers a resistance to desiccation and helps the organism to survive in the environment. The drawback is that hydrophilic nutrient uptake is going to be that more difficult than

• **Figure 1.11** The cell wall structure of Mycobacteria. Note how the mycobacterial cell wall is dominated by the fatty acid mycolic acid which combines with various carbohydrates to form the thick, waxy layer. The mycolate structures make the cell wall relatively impermeable to water, and hence resistant to drying

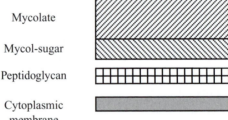

Mycolate

Mycol-sugar

Peptidoglycan

Cytoplasmic
membrane

---

### ■ BOX 1.5 THE ZIEHL–NEILSON (ZN) STAIN

Having prepared a slide with a smear of the suspected Mycobacterium *sp.* and fixed the slide by passing it briefly through a Bunsen burner flame, the ZN stain is carried out as follows:

Carbol fuschin (0.3 per cent w/v). Heat slide on a hot plate and leave stain for 3–5 mins once steaming starts.

3 per cent hydrochloric acid in alcohol. Decolourise until no visible red stain leaves the slide.

Malachite green (0.5 per cent w/v) 1 min.

Each step is followed by rinsing the slide with water. Blot dry and view using the ×100 objective with immersion oil to concentrate the light into the objective.

Mycobacteria stain red, non-acid fast bacteria stain green.

---

Gram positive bacteria. The limited flow of hydrophilic compounds across the mycobacterial cell wall will be via porins.

## ■ 1.2.2 CELL MEMBRANE

Unlike the cell wall, if the cell membrane is disrupted the cell dies as a result of osmosis: the intracellular ions rapidly exit whilst external water flows into the cytosol. Although situated beneath the peptidoglycan layer, the hydrophobic phospholipid membrane called the cytoplasmic membrane acts as the barrier protecting the cytosol from the outside world (at least in Gram positive organisms). The cytoplasmic membrane in most microbes has a bilayer structure similar to that of eukaryotic membranes but differs in the chemical composition. Mammalian membranes are stabilised by a high proportion of sterols such as cholesterol, whereas sterols are absent from prokaryotic membranes. Instead, in those bacteria that have them, compounds called hopanoids are thought to play a similar role. Hydrophilic intracellular solutes need to be transported across the cytoplasmic membrane and then they are able to freely diffuse through the hydrophilic peptidoglycan. With Gram negative bacteria the lipopolysaccharide is the limiting face of the cell to the external environment and movement of hydrophilic solutes occurs via porins. Whereas many small molecular size nutrients will pass down a concentration gradient from the outside (high concentration) to the inside of the bacterial cell (low concentration) by diffusion, we need to consider the export of proteins that are needed outside the cell. Many proteins are manufactured for functions that occur outside the cell such as hydrolytic proteases, lipases and carbohydrate-splitting enzymes. The trans-location of these large molecular weight proteins across the cytoplasmic membrane occurs mostly by an active secretory mechanism involving the Sec pathway known as the **general secretory pathway**. The general secretory pathway utilises ATP to drive a complicated membrane transport process in which the protein is translocated by various Sec proteins. Sec proteins act as molecular chaperones, escorting the protein on its path through the membrane, having recognised a particular stretch of amino acids on the amino terminal end of the protein (called the **signal sequence**). The translocation across the cytoplasmic membrane is termed **protein secretion**. With Gram positive bacteria secreted proteins can be detected in the culture medium and this net effect is called **excretion**. For Gram negative bacteria the secreted proteins will end up in the periplasmic space, held between the cytoplasmic membrane and the outer membrane, such that the proteins will not appear in the culture medium (without the lysis of the cell). The transport of proteins across the outer membrane can occur by several different mechanisms: **secretion systems types I, II, III and IV**. Types I and III do not employ Sec proteins and type II secretion is the most widely used system. Type III secretion is discussed further in Chapter 8 as toxic proteins are delivered into host cells by this mechanism.

Movement of proteins across bacterial membranes often involves specific proteins to assist the process, typically Sec proteins. They act as molecular chaperones, stabilising the transported protein through the membranes.

As with eukaryotic cells, the cytoplasmic membrane is a site of considerable activity in terms of signalling and exchange of materials (Box 1.6). For bacteria the absence of specialised internal membranes means that the cytoplasmic membrane is also the site of respiratory chain and oxidative phosphorylation. The cytoplasmic membrane is a dynamic structure, undergoing changes in its composition in response to the changing environment the bacterium encounters.

## ■ 1.2.3 NUCLEIC ACID

In general, bacteria possess a single closed circular loop of DNA as their chromosome. The bacterial chromosome is traditionally considered to be the DNA that contains the

---

### ■ BOX 1.6 FUNCTIONS OF THE CYTOPLASMIC MEMBRANE

Cytoplasmic membrane functions include:

- osmotic barrier,
- transport of solutes (nutrients and ions),
- exclusion of charged toxic compounds,
- assembly and transport of peptidoglycan,
- energetics: the electron transport chain,
- site of signalling with extracellular ligands.

---

essential information for growth and replication. The chromosome is not a fixed isolated unit but under regular assault from other mobile genetic elements that may insert or remove genes. These mobile genetic elements, which include **bacteriophages** and **plasmids** (see pp. 31–35), are often only temporary passengers. The constraints of size will mean that rate of gene accumulation and gene loss will be approximately equal.

Stressing that bacteria have a single chromosome is the general rule, there are plenty of exceptions. For example *Vibrio cholerae* has two chromosomes. All bacteria, however, have only one copy per cell (**haploid**), unlike the multiple linear chromosomes of mammalian cells. Some bacteria have linear chromosomes but the advantages of such are not clear (the ends of linear chromosomes are prone to enzymic attack). The defining feature of prokaryotes is that there is no membrane surrounding the nucleic acid, unlike eukaryotes. Furthermore, in contrast to eukaryotes, bacterial DNA is not bound by basic, histone-like proteins thereby permitting rapid transcription and translocation of genes. Of the increasing number of bacterial genomes that have been sequenced, the size appears to be around $5 \times 10^6$ base pairs, with *Escherichia coli* having approximately 4000 genes. Unlike eukaryotes, bacterial DNA (mostly) does not contain non-coding sequences of DNA (**introns**); that is to say, the entire genome is functional.

One of the most intriguing features of the bacterial cell is the fact that the nucleic acid is able to fit inside it at all. The extraction of the nucleic acid from bacteria will result in the nucleic acid denaturing and unravelling. The length of the unravelled DNA is around 1 mm, something like a thousand times the length of a coccus measuring 1 μm in diameter. This engineering feat of packaging the DNA into the cell is achieved through repeated coiling of the DNA strand (Box 1.7. and Figure 1.12). This 'supercoiling' can be visualised to a degree by twisting taut string in opposite directions with the subsequent formation of tight coils resembling a twisted telephone cable.

The enzymes responsible for catalysing the supercoiling of DNA are called **gyrases** and the controlled unwinding of the supercoil is mediated by **topoisomerases**. The use of enzymes tells us that the processes are tightly controlled events. Random bunching of the nucleic acid would be too chaotic and hinder efficient replication and transcription.

---

### ■ BOX 1.7 THE 3 PER CENT KOH TEST

Denatured nucleic acid from bacteria can be visualised by this simple test using Gram negative bacteria mixed with 3 per cent potassium hydroxide. The KOH dissolves the cell wall and the denatured nucleic acid forms a gel-like substance that can be seen attaching to a wire loop lifted from the suspension of cells.

---

• **Figure 1.12** The ordered reduction in size of the bacterial chromosome is achieved through appropriate enzymes. The individual loops of DNA resemble a series of supercoiled circular plasmids. This arrangement helps the location of specific sites in the DNA which facilitate transcription and replication events

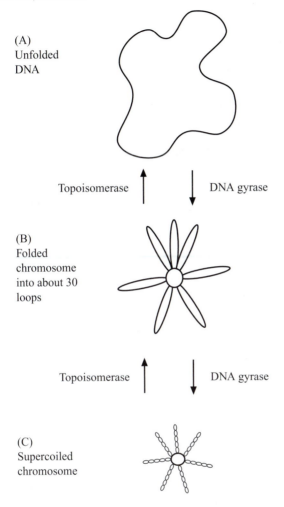

(A)
Unfolded
DNA

Topoisomerase    DNA gyrase

(B)
Folded
chromosome
into about 30
loops

Topoisomerase    DNA gyrase

(C)
Supercoiled
chromosome

Remember, some bacteria can divide every 30 minutes! Whilst not histones, certain proteins associated with the DNA are detectable and are thought to help with the supercoiling.

The speed of transcription and translation depends in part on the efficiency with which the appropriate genes on the DNA are temporarily uncoiled from the supercoiled state so as to start transcription. The tendency of DNA to be arranged in operons (clusters of genes of related function) along with **haploidy** (a single copy of the chromosome) promotes a simple control of such operons (any more than one copy of a gene will complicate its control).

It is well known that the four bases that comprise the DNA occur in pairs. Adenine (A) pairs with thymine (T) and guanine (G) with cytosine (C) so that these base pairs occur in equal amounts per cell. The proportion of the total DNA that is G+C varies considerably between different bacterial genera and within members of the same species, but less so than between genera. The G+C content (expressed as G+C mol per cent) is thus a useful means of distinguishing between bacterial species. For example, a difference of 15 mol per cent G+C between two bacterial strains of the same genus indicates that they are separate species. Information on the entire gene sequences for different bacteria is appearing in the scientific press with increasing frequency as the methods are becoming automated. Of the numerous interesting features being uncovered, the variation in gene sequence between different strains of the same species can vary by up to 20 per cent! The variation is in total genome size as well as the sequence variation. For any particular bacterial species, the genes in the chromosome are classed under two sets:

- the core set of genes, and
- auxiliary genes.

The core set will comprise those that code for the essential characteristic features of the species in question and the auxiliary genes will code for the properties that are not found in all strains of that species. For example, utilisation of the carbohydrate sorbitol is found in between 10–40 per cent of strains of *Escherichia coli*. Auxiliary genes account for the reason that some strains of *Escherichia coli* cause disease in humans whereas others do not. The variability in the gene pool in the organism can account for the exploitation of new environments by bacterial strains. Although termed auxiliary, both core and auxiliary genes are under selective pressure such that redundant genes will eventually be lost.

In many bacteria additional lengths of DNA can be found, separate from the chromosome. Called **plasmids**, these independently replicating circular loops of DNA code for proteins involved in a wide variety of functions. These include production of bacteriocins, virulence factors, degradative enzymes and resistance to antibiotics. Plasmids are superfluous to the normal replication of the bacterium but will clearly offer a selective advantage when, for example, the organism is exposed to antibiotics. Plasmids may incorporate into the chromosome (when they are called 'episomes') and can confer the genetic information necessary for conjugative transfer, i.e. the production of the sex pili by the F plasmid through which the plasmid is transferred to another bacterium. Further information is given below (p. 31).

### ■ 1.2.4 RIBOSOMES

The ribosome is a collection of proteins and different ribosomal RNA molecules that form the site of protein synthesis (Figure 1.13a).

Bacterial protein synthesis can be considered a streamlined version of that seen in eukaryotic cells:

- smaller ribosomes need fewer start signals (initiation factors),
- bacterial mRNA is polycistronic (see Figure 1.13b and Figure 1.18),
- translation of mRNA occurs at the same time as the mRNA is being manufactured,
- the absence of a nuclear membrane obviates the need for movement of mRNA from nucleus to cytoplasm.

All these factors contribute to a faster rate of protein synthesis.

The differences that exist between prokaryotic and eukaryotic ribosomes can be exploited as antibiotic targets. The differences include size of the protein subunits: prokaryotes have lower molecular weights than eukaryote subunits and these are seen in different sedimentation units (S) when centrifuged. Also, bacterial ribosomes exist as multiple functioning units all catalysing translation whilst attached to a single RNA

• **Figure 1.13** The ribosome. (A) Diagrammatic representation of the bacterial 70S ribosome and its components. (B) The production of polycistronic mRNA permits the binding of multiple ribosomes to form the polysome

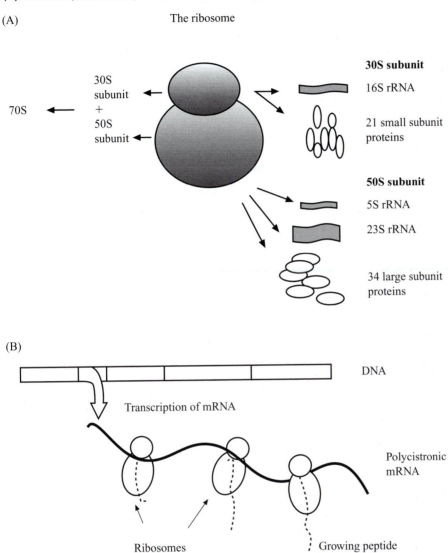

(A)

The ribosome

30S subunit

16S rRNA

21 small subunit proteins

50S subunit

5S rRNA

23S rRNA

34 large subunit proteins

70S

30S subunit
+
50S subunit

(B)

DNA

Transcription of mRNA

Polycistronic mRNA

Ribosomes

Growing peptide chains

strand (Figure 1.13b). The resemblance to a string of beads on a thread is a useful (and widely used) image.

### ■ 1.2.5 STORAGE GRANULES AND INCLUSIONS

In some species of bacteria it is possible to detect intracytoplasmic granules of reserve materials called 'storage granules' or 'inclusion bodies'. Their presence can be demonstrated with various stains directed at the composition of the storage granule. Polysaccharide granules are usually starch or glycogen and therefore stained blue or brown respectively with iodine. Lipid granules are often polyhydroxybutyrate and are stained by lipophilic dyes such as Sudan black. The other type of storage granule is that found in *Corynebacterium spp.*, known as volutin (or metachromatic) granules which stain with aged methylene blue dye. They are accumulated phosphate groups and, like all the above storage granules, represent a valuable energy store.

### ■ 1.2.6 SURFACE STRUCTURES: S-LAYERS

S-layers (paracrystalline surface layers) are porous, proteinaceous layers on the bacterial cell surface. Where they have been described, and this includes all archaebacteria and a large number of human and animal pathogens, their function differs. As an external covering for the bacterium it might not be surprising that they have been implicated in protection from phagocytosis, adhesion to surfaces and as a variable antigenic coat. The significance of S-layers has yet to be fully understood.

### ■ 1.2.7 CAPSULE

Capsules and slime layers are extracellular substances, usually polysaccharide polymers, which are not essential for the continued replication of the organism. As a loosely attached coat, capsules may be lost, but their presence offers distinct advantages under certain circumstances. As with S-layers, the extracellular location confers properties such as adhesion to substrata (Figure 1.6), protection from phagocytosis and antibody action. Early studies in mouse models demonstrated that the virulent strains of pneumococci were capsulate, unlike the avirulent non-capsulate strains. Capsules are not only involved in the infectious process, they are also thought to offer protection against desiccation through acting as a reservoir for water. Capsules can be visualised easily by using stains such as nigrosin and India ink. These stains do not actually penetrate the capsule but simply create a black background against which the colourless capsule is seen surrounding the vegetative cell. For this reason this technique is known as a 'negative stain'. Antibodies raised against capsules in *Escherichia coli* and other members of the Enterobacteriaceae (the family encompassing the fermentative enteric Gram negative rods) are given the prefix K (from the German word 'Kapsule'). *Escherichia coli* K12 is one of the most well-characterised bacteria.

### ■ 1.2.8 FLAGELLA

Flagella (singlular: flagellum) are hollow cylinders of a fixed helical shape that are attached to the body of the bacterium (the vegetative cell). They provide a means of directed movement for the bacterium. They are mounted on a rotating plate that will spin the flagellum through an energetically driven process (utilising proton or sodium gradients between the inside and outside of the cell) at speeds of around 15,000 rpm! The flagella all rotate in synchrony for brief periods, thus projecting the organism in one direction. Flagella are usually more than twice the length of the vegetative cell and composed of protein subunits (called 'flagellin'). Being protein in nature, they are immunogenic to man and the presence of antibodies to flagellar antigens (termed 'H antigens') can be detected in several infections. Their number and position on the cell provides useful distinguishing features (Figure 1.14). However, not all bacteria possess flagella. Particles greater then 4 μm in size will sediment in suspension whereas those smaller than this will be kept in suspension through Brownian motion. This will give ample opportunity for bacteria to be carried by movement of the solution it finds itself in.

### ■ 1.2.9 FIMBRIAE/ADHESINS

**Fimbriae** appear in electron micrographs as fine hair-like projections radiating from bacterial cells. They mediate the attachment of the bacterium to a surface or other cells. They are constructed from protein subunits, which are added to the fimbrial base

• **Figure 1.14** Some patterns of flagella arrangement in bacterial rods

Monopolar
monotrichous

Monopolar
polytrichous

Peritrichous

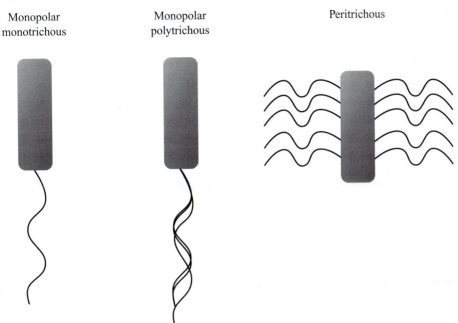

to give a helical pattern. They can be up to $20\,\mu m$ in length and vary in their number per cell. The adhesion of the fimbriae to their site attachment is due to a specific receptor that is only expressed at the tips of the fimbriae. These adhesive sites are specific to their appropriate receptors rather than a general non-selective adhesive paint. The placement of the adhesive component on the end of a stalk is suggested to help overcome charge repulsion between the bacteria and the surface.

When bacteria that require oxygen for growth (aerophilic) form a dense layer of cells on the surface of a fluid (called a 'pellicle') fimbriae are thought to help maintain gaseous exchange across individual cells by keeping bacteria apart.

Similar to fimbriae in appearance are **sex pili** (singular: pilus). These structures are involved in the exchange of genetic material between two bacterial cells known as 'conjugation'. Whilst the terms 'fimbriae' and 'pili' are often used interchangeably, it is recommended that pili be reserved for those structures involved in conjugation.

Bacteria probably spend the majority of their time as large populations of cells rather than as solitary cells. The adhesion of bacteria to surfaces and each other create masses of cells termed **biofilms**. Such large populations of bacteria are not confined to the crevices of our gums or intestines but can be felt covering rocks in streams. The ability to resist flushing will be of clear benefit in the course of establishing a population of bacteria in the river or urinary tract.

Adhesive properties do not only reside in fimbriae. Certain bacteria express coats of protein that act as **adhesins** (e.g. M-protein in *Streptococcus pyogenes* is important in aiding adherence of this organism to the epithelia prior to causing tonsillitis in humans).

### ■ 1.2.10 ENDOSPORES

As a means of providing a dormant but robust means of survival, sporulation (the production of an endospore) is highly effective. They are produced in response to nutrient limitation and once the **vegetative** cell (the part of the bacterial cell that is not the spore is called the 'vegetative cell' or 'sporangium') has lysed the endospore will persist in an inactive state until certain triggers activate the spore such that a new vegetative cell emerges (**germinates**). Not all bacteria have the capacity to produce spores and neither are they essential for reproduction in those species that can produce them.

Production of spores is a distinguishing feature of the genera *Clostridium* and *Bacillus*. The shape of the spores and their location within the cell is of some value in identifying the organism (Figure 1.15). Spores are oval or round and, as a general rule, are often wider in clostridia in contrast to those in the genus *Bacillus*. They do not stain with Gram stain and hence appear as colourless structures. Special staining procedures are used to drive dyes into the spore wall in order to visualise them.

Their structure consists of a series of tough cell wall layers (germ cell wall, cortex, spore coat) surrounding an inner core that contains the nucleic acid and protein essential for outgrowth at a later stage. The spores are dehydrated and withstand boiling, chemical treatments and radiation that have formed the basis for sterilisation procedures in food and medical products (see Chapter 5). Revival of dormant spores can be divided into three steps: activation of the core, germination (the start of metabolic activity and rehydration) followed by outgrowth of the new vegetative cell.

## ■ 1.3 BACTERIAL MOVEMENT

Being small has implications for the mobility of micro-organisms. Many organisms are transmitted through the air. Fungal spores create specialised structures to ensure that the spores are sufficiently above ground level to get carried in the wind. Being small, spores and bacterial cells remain suspended in air for long periods. What are the effects of size on mobility in fluids? In fluids, the movement of particles is dependent on size. Whereas larger particles will fall to the bottom, smaller particles (<5 μm) will be kept suspended due to Brownian motion. Clumps of bacteria will therefore sediment. Bacteria are able to sense gradients in chemicals and respond chemotactically. Chemotaxis requires directional movement towards nutrients or away from toxic chemicals. The use of flagella provides the mobility in fluids and the only example of the use of a rotor mechanism in biology. The reason that bacteria invented the wheel lies in the need to move small particles through viscous fluids. Large organisms can swim because the

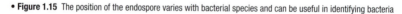

• **Figure 1.15** The position of the endospore varies with bacterial species and can be useful in identifying bacteria

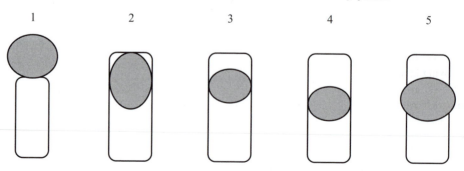

1: terminal (drumstick)
2: terminal (oval spore)
3: subterminal
4, 5: central
Note how the spore width can exceed that of the vegetative cell

inertia provided by their mass propels them through a fluid sufficiently to enable the return stroke of their fins, arms or tails. Micro-organisms do not possess sufficient mass to provide any movement to overcome the forces of inertia. Being so small they only have to overcome the viscosity of the fluid (forces required to accelerate masses). In other words, when bacteria stop swimming by flagella they come to an immediate stop. The relationship between inertia and viscosity is given by a formula established by Osborne Reynolds (**Reynolds number**). Bacteria have very small Reynolds numbers and have only to deal with viscosity, unlike humans (and at the opposite extreme whales!) which have to overcome inertial forces. If bacteria had fins or arms a forward stroke would propel the organism forward equivalent to the volume of water displaced *but* upon the return stroke the organism would move the same distance in the reverse direction. The organism would not move anywhere but, instead, oscillate between the distance of the forward and backwards propulsion. The reason is that the bacterium has insufficient inertia to overcome viscosity. To overcome this, bacteria use a rotating flagellum. In this way the organisms can continually rotate the flagella without the need for the equivalent of a 'return stroke'. For spiral-shaped bacteria the helical-shaped flagella within the body of the organism and the rotation of the spiral propels the organism forward. Again the rotation in one direction does away with the need for any 'return stroke'.

Figure 1.16 shows that when the flagella all rotate together they form a helical shape which projects the organism forward. To change direction the organism reverses the rotation of the flagella, the bundle comes apart and each flagellum rotates independently and the directional force is lost. The organism then tumbles randomly until the flagella all reverse rotation and form the bundle again. The ability to control the direction in which the organism moves is controlled by the frequency with which the organism 'swims' or tumbles. If the organism senses an increasing gradient of nutrients then the organism keeps swimming, whereas if the gradient is of a noxious substance then tumbling is invoked.

It is worth noting that flagella are rare in spherical bacteria. Theoretically, calculations indicate that the optimal shape for bacteria movement through fluids is a cylinder. For rod-shaped bacteria, an optimal SA/V ratio can be maintained with varying length but also the appropriate length to width ratio for the movement can be achieved. Thus spherical objects (cocci) offer too much resistance to movement compared with cylinders. Motility in cocci instead occurs by a mechanism called 'twitching'. The retraction of pili that are attached to a surface is used to pull the organism along the surface.

### ■ 1.3.1 PLANKTONIC OR SESSILE?

When cultured in the laboratory, attention is generally focussed on the bacteria that are growing in suspension rather than those that attach to the wall of the container that the bacteria are growing in. Those organisms that drift or float in suspension are called **planktonic**. This may be appropriate for bacteria in any fluid medium, be it oceans and lakes or saliva. However, it is increasingly likely that most bacteria spend most of their existence as **sessile** organisms: adherent or attached to surfaces. Bacteria in nature will tend to adhere to surfaces, not least because in solutions of very low nutrient content the electrostatic charges (mostly negative) that exist on surfaces will attract and concentrate nutrients as a layer facing the surface. This concentrating effect can be exploited by the bacterial **biofilm** that develops on surfaces. The film of organisms develops quickly, the nutrient concentration falls and the growth rate falls correspondingly. Bacteria that are nutritionally starved and are in stationary phase are generally

Bacterial flagella drive chemotaxis. (A) The bacterial cell will tumble randomly when the flagella are splayed but (B) e flagella form a single helical, corkscrew shape that drives the cell forward. (C) shows how the frequency of the creases as the cell senses the increasing concentration of the attractant

fumbling

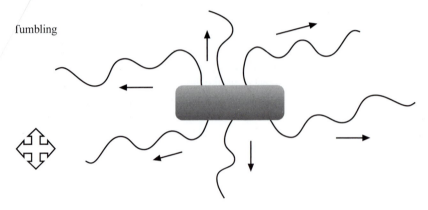

(B) Swimming

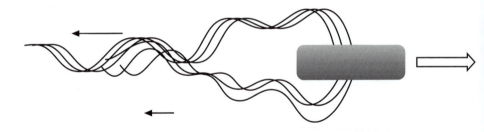

(C) Chemotaxis

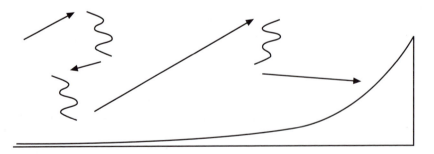

Gradient of attractant

more resistant to the antibacterial action of antibiotics, disinfectants such as chlorine and heavy metals. This relative resistance is aided by the physical barrier of bacteria and secreted exopolysaccharide, capsular material (slime) which, in addition, may limit desiccation of the biofilm. These benefits may be available to bacteria that colonise inert surfaces in humans such as infected catheter lines and intravenous cannulae. Treatment with antibiotics will remain ineffective until the colonised tube is removed from the patient.

    **Biofilms** are common modes of existence for populations of different species of bacteria. This need not represent competition for the different species present. In the plaque layers that can accumulate on teeth the excreted by-products of different species are used by other bacterial species for nutrients. Likewise, depending on the oxygen tolerance, different organisms position themselves according to their gaseous requirement. Clearly, benefits arise from growing as a population of cells, either in a multi-species environment or as a population of a single species.

> The term 'biofilm' is used to describe normal and abnormal accumulations of bacteria. The accumulation of bacteria on surfaces or in equipment in industrial settings is invariably an unwanted problem, whereas bacteria adherent to the mucous membranes in the body is the normal mode of existence.

### ■ 1.3.2 QUORUM SENSING: BACTERIA AS MULTICELLULAR ORGANISMS

In contrast to viewing bacteria as individual cells in competition with every other organism in the environment, bacteria (of the same species) have been shown to act in concert by signalling to each other and co-ordinating their behaviour. The communication system is called **quorum sensing** after the Latin to indicate that sufficient members are present for voting. The signals used fall broadly into two types, Gram positives tend to use small peptides whereas Gram negative bacteria use **N-acyl-homoserine lactones** (AHLs). Taking an organism such as *Escherichia coli*, quorum sensing works by all members of the population releasing the signalling lactone into the environment. Once a threshold concentration of the lactone is reached, it is able to trigger the regulation (up-regulation or down-regulation) of gene expression in the population of the cells as appropriate (Figure 1.17). The signal is an **autoinducer**, as it is produced by and works on itself and other cells in the growing population. The effects may be to enter stationary phase of growth once a suitable density of cells in the biofilm has been reached. Other effects include the regulation of expression of virulence factors such as exotoxins and degradative enzymes once inside the new host. By using quorum sensing, the organism prevents unnecessary activity until the appropriate time. By producing a virulence factor, for example only when large numbers of the organism are present, the total amount of virulence factor is optimised and hence the bacteria stand a greater chance of having an effect. Consequently, this cell density-regulated gene expression appropriately has been viewed as a potential target for antibiotics.

### ■ 1.4 BACTERIAL GENETICS

The bacterial chromosome is of considerable molecular weight and varies roughly tenfold, i.e. $10^6$ to $10^7$ base pairs per cell with individual genes being roughly 1000 base pairs each. *Escherichia coli* has around 4000 genes. Bacteria generally have just the one copy but, as more chromosomes are sequenced, organisms are increasingly found to possess more than one chromosome. Even if there is more than one chromosome, bacteria are still **haploid** and one of the consequences is that recessive mutations are always expressed in the bacterial phenotype, in contrast to diploid organisms.

    The term **genome** refers then to the entire collection of genes within a cell, be they located in a chromosome or plasmid. The **genotype** refers to the total genetic information of a cell whereas the **phenotype** refers to the observed properties. The size of the genome reflects the natural habitat: extracellular or intracellular existence.

(The term 'logarithmic growth' is also used but it is not sufficiently accurate as the type of logarithm is not specified.)

**Why is it called exponential growth?** The numbers of organisms increases according to a mathematical formula that uses exponents and therefore is called exponential growth. The important feature of exponential growth is that the rate of increase in bacterial numbers is proportional to the number of organisms at that time. In other words, the more bacteria there are, the greater the increase in their numbers. Figure 2.1 illustrates the differences between exponential and power laws.

The numbers of organisms that theoretically are produced during growth is given in Table 2.1. In the last column, the increase in bacterial numbers is given in mathematical terms and $n$ is an exponent. The number of organisms that are present after $n$ generations will be $2^n \times N_0$. If the original number of organisms was one, then the number of organisms after $n$ generations will be $2^n \times 1$. Note, then, that the increase in the number of organisms will be dominated by the value of the exponent. If the original number of organisms was greater than one, i.e. $N_0$, then the formula will be

$$N_0 \times 2^n \text{ (or } 2^n N_0) \tag{2.1}$$

For a culture of *Esch. coli* that doubles every 30 minutes, if we start with six organisms, after eight hours there will be: $6 \times 2^8$ (1536, in case you were not sure). The consequence of multiplying the original number of bacteria by a number that increases according to the $2^n$ is that the increase is not linear. This increase is not in direct proportion to time taken (i.e. a linear increase) but, instead, increases geometrically. A geometric increase is obtained when the difference between successive numbers increases by a ratio ($2^n$) every time (1, 2, 4, 8, 16, 32, 64 . . .). With arithmetic progressions the difference between the successive numbers is fixed, i.e. will be the same every time (e.g. 1, 2, 3, 4, 5, 6 . . .). The rate of increase is constant in arithmetic progression, whereas geometric increases rise exponentially.

If the numbers of organisms are plotted against time using ordinary graph paper the results very quickly become impossible to plot because the numbers of organisms rises so quickly to very large values. The graph paper is simply not big enough to fit the numbers of organisms on. It is therefore necessary to plot the logarithm values of the numbers of organisms on the $y$-axis instead. The $x$-axis is plotted normally and the result is termed a 'semilogarithmic plot'. The use of logarithms enables one to handle large numbers in simple terms. You recall that logarithms express numbers as exponents of a base.

**Table 2.1  Increase in numbers of bacteria following binary fission**

| Number of generations* | Numbers (total) | Numbers (expressed as exponents) |
|---|---|---|
| 0 | 1 | $2^0$ |
| 1 | 2 | $2^1$ |
| 2 | 4 | $2^2$ |
| 3 | 8 | $2^3$ |
| 4 | 16 | $2^4$ |
| $n$ | $2^n$ | $2^n$ |

* One bacterial cell multiplied by the number of generations.

• **Figure 2.1** Comparison of power functions and exponential functions. (A) and (B) show how power functions can result in positive and negative numbers. (C) illustrates the pattern of exponential functions. (D) illustrates the section from (C) where values less than 1 approach but do not reach zero

Power functions

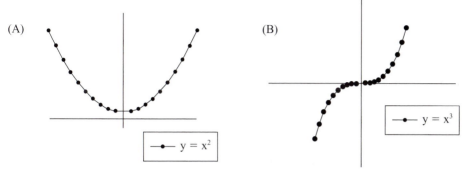

Exponential functions

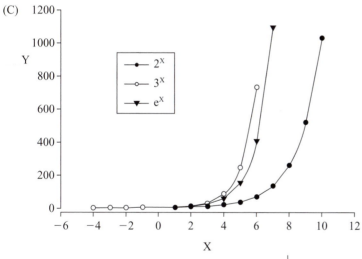

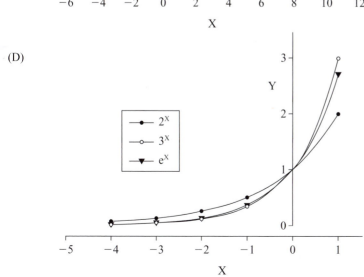

• **Figure 2.3** Synchronous growth. The solid line shows the doubling in bacterial numbers if all the bacteria divide in a synchronised step (had the same generation time). In practice, the small variations in generation time for each bacterium means that the numbers increase without such steps, and the slope describes the increase in bacterial mass and numbers

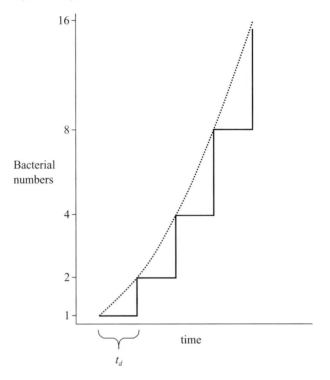

If different organisms have different growth rates under unrestricted conditions then the rate of increase in both cell number or in cell protein or nitrogen can be described numerically. This term is called the **specific growth rate** and the symbol $\mu$ is used (but, confusingly, the letter k is used in certain texts and instead of specific growth rate constant it is also called the 'growth rate constant'). $\mu$ represents the *instantaneous* rate of increase of a single cell rather than the increase measured at fixed intervals. Thus, $\mu$ is distinct from doubling times. We have seen that the exponential increase in growth means that the rate of increase is directly proportional to the cell mass at that point in time. When using bacterial mass (protein, weight, etc.) to measure exponential growth, the increase at any one point in time represents $\mu$. It is derived from the mathematical expression of the rate of increase in cell growth. This increase will be over time (i.e. a rate) and is:

$$\frac{dN}{dt} = \mu N \tag{2.3}$$

With unrestricted conditions for growth (appropriate type and quantity of nutrients), an organism will grow at its maximum specific growth rate ($\mu_{max}$). Such conditions do not last long in batch culture.

This expression says that the rate of increase in bacterial numbers (dN) over time (dt) is equal to the specific growth rate ($\mu$) multiplied by the number of organisms. In other words, the increase in numbers goes up proportionally with the number of organisms present: the more organisms there are, the faster the numbers increase. This constant of proportionality, $\mu$, represents the number of doublings per unit time (usually per hour) as well as describing a property of the bacterial population (how fast they can grow). The value of $\mu$ is its use as a marker of the response of a bacterial culture to changes in its environment. Alterations in growth rate can be measured almost immediately following changes in the culture conditions and provide a sensitive indicator of the physiology of the cell.

Equation 2.3 does not specify any particular time. The population of bacteria at any one point in time will have a mixture of bacteria at different stages in their division. To calculate the slope of a curve is virtually impossible with rulers on graph paper so the rate of increase is calculated by integration.

Those of you who are familiar with integration can skip to equation 2.4. Others may need to take the process one step at a time. For any one point in time equation 2.3 is integrated as follows.

Rearrange the equation to put all terms concerning bacterial numbers (N) on the same side:

$$\frac{dN}{N} = \mu dt$$

And then separate one side into separate components:

$$\frac{1}{N} \cdot dN = \mu dt$$

You may recall that the integration of $\frac{1}{N} \cdot dN = \ln (N)$, hence we can integrate both sides between time $t_0$ and $t_1$.

$$\ln (Nt_1) - \ln (Nt_0) = \mu (t_1 - t_0)$$

Take exponentials of both sides:

$$e^{\ln (Nt_1)} \cdot e^{-\ln (Nt_0)} = e^{\mu (t_1 - t_0)}$$

This cumbersome formula can be simplified because of two mathematical rules:

$$e^{\ln (x)} = x$$

and

$$e^{-\ln (x)} = 1/x$$

Hence:

$$Nt_1 \cdot \frac{1}{Nt_0} = e^{\mu (t_1 - t_0)} \qquad (2.4)$$

Or

$$\frac{Nt_1}{Nt_0} = e^{\mu (t_1 - t_0)} \qquad (2.5)$$

In summary (see also Box 2.1), two mathematical methods have been described to examine balanced bacterial growth. One is described for counting bacterial numbers at a defined time period (equation 2.1) and the other is used to examine growth at a point in time (instantaneously) using equation 2.6. These mathematical exercises have arisen because the bacterial population when multiplying vigorously will increase in an exponential fashion. The actual numbers of organisms can be plotted against time using $\log_2$ for the $y$-axis as this is a logical description of the increase in bacterial numbers. The time taken to double the numbers is the doubling time.

## ■ 2.2 THE GROWTH CURVE: GROWTH IN BATCH CULTURE

When bacteria are inoculated into a fresh bottle of growth medium and followed over a period of time without replacing any of the culture, a typical pattern of growth is observed. The typical shape of the growth curve is seen in Figure 2.4 in which the viable count and *not* the total count of bacteria is plotted. As the organisms adapt to the new environment, the numbers do not significantly alter. This is called the **lag phase**. Once cells have adjusted and start to multiply, the numbers rise exponentially (**the exponential phase**). This rise in bacterial number will continue until the nutrients become depleted or limiting and the toxic waste products of metabolism reach an inhibitory point. The increase will diminish until it levels off. This plateau is called the **stationary phase**, where the increase in bacterial numbers is balanced by the increasing number dying. Whilst the numbers of organisms remains steady, the term stationary does not describe the metabolic state of the organisms. The bacteria that are alive at this point are still actively metabolising what little substrate remains. Eventually, the environment is depleted of one or more essential nutrients and the organisms are being killed by the toxic environment, usually due to the extreme pH which usually becomes very acidic. This last stage where numbers fall is called the **decline phase**.

---

### ■ BOX 2.1 BACTERIAL GROWTH: THE TERMS

As bacteria grow, they increase their biomass and their numbers. In cultures it is likely that growth is unbalanced; that is, increase in cell size does not occur in all organisms simultaneously, therefore the increase in cell numbers as they all divide is likewise not simultaneous. Thus, increase in cell size and cell numbers are not the same at any one point in time.

As a population of bacteria grow and divide, the growth is exponential, hence the **growth rate** (per unit of time) changes with time (getting faster as the numbers increase). The **specific growth rate** ($\mu$, measured as $hr^{-1}$), however, is a constant and therefore fixed during exponential growth.

The **generation time**, measured in hours, represents the time taken for a cell to replicate. The **mean generation time** (g) is the time taken by a population of cells to increase in bacterial numbers.

The **doubling time** (td) is the time taken for a culture to double in numbers or biomass. In theory, it should be the same period of time as taken for all the cells to double (the mean generation time) if *all* cells were dividing. This need not be the case. (Consider 100 cells all taking 2 minutes to divide. In 10 minutes there will be 3200 cells. If 100 cells stop dividing after 5 minutes then the time taken to reach 3200 bacteria will be 12 minutes. The mean generation time for the growing population will be 2 minutes and the doubling time will be the same, but the results hide the fact that some cells have stopped dividing.)

• **Figure 2.4** Typical phases of bacterial growth obtained in batch culture. (A) shows the untransformed data, in which it is difficult to determine the slope of the rate of increase in bacterial numbers. In (B), however, by plotting the natural log of the bacterial numbers the slope of the exponential phase can more easily be determined. Whilst this general pattern is obtained in batch culture for most organisms, the time spent in each phase will vary between organisms and between different growth conditions (temperature, culture media, etc.)

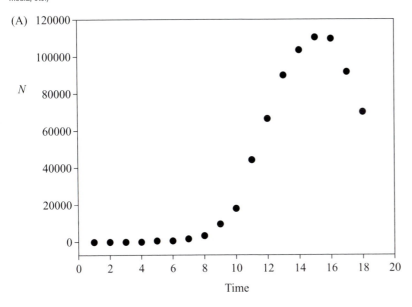

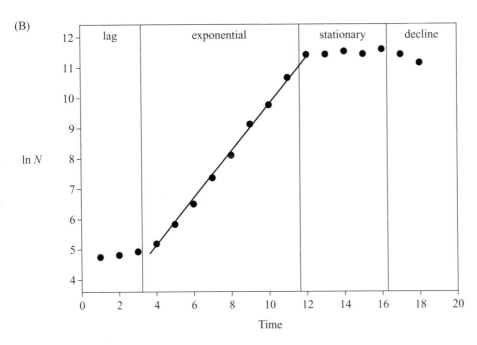

source (glucose) but, after a limited period, the growth rate increases again. The explanation is the activation of the *lac* operon *after* the glucose has been fully utilised. The relationship between concentration of substrate and growth rate has been widely studied in many organisms. One of the key principles was established by Jacques Monod and colleagues. Figure 2.8 shows how the concentration of substrate and growth rate is *not* a linear function but instead has a hyperbolic relationship, which is best described by the **Monod equation**. The value of the parameters such as $\frac{1}{2}\mu_{max}$ and $K_S$ is then seen in Figure 2.9 where the competition between organisms is predicted from the $K_S$ for each.

## ■ 2.4 THE PHYSICOCHEMICAL PARAMETERS THAT INFLUENCE BACTERIAL GROWTH

### 2.4.1 TEMPERATURE

The bacteria that colonise or infect man are **mesophiles**; that is, have optimal growth temperatures between 20 and 40°C (Figure 2.10). **Thermophiles** are those organisms that grow at elevated temperatures (a good example being those found in thermal lakes) and are not known to infect man. **Psychrophiles** are those that grow at reduced temperatures below 20°C. These labels are not mutually exclusive. *Listeria monocytogenes*, for example, is a mesophilic organism that can cause infections in man, but can grow at 4°C, that is, in a fridge!

Whilst temperatures below 20°C will inhibit the growth of most human pathogens, reduced temperatures will not kill bacteria (either the vegetative cell or the spore), viruses or fungi. Conversely, raising the temperature generally accelerates growth rates, unless the change is abrupt, when it is detrimental to the viability of the organism. Increasing the temperature will only increase growth up to a certain point, above which it is likely to become suboptimal, if not lethal (Figure 2.11).

• **Figure 2.8** Effect of substrate concentration [S] on growth rate ($\mu$). The substrate is growth limiting, but this only has a significant effect in reducing growth rate at relatively low concentrations. The curve is a hyperbolic (resembling rectangular) shape and is described by the Monod equation:

$$\mu = \mu_{max}\, (S/(K_S + S))$$

where $\mu_{max}$ is the maximum growth rate at unlimited substrate concentration and $K_S$ is the concentration of substrate that yields half maximum growth rate

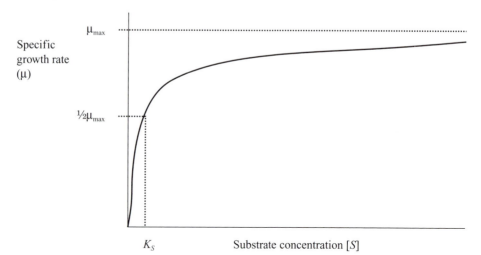

• **Figure 2.9** Effect of substrate concentration on competition between two organisms. When organisms have different affinities for the substrate ($K_{S1}$ and $K_{S2}$) and different maximum growth rates ($\mu_{max}$) the outcomes will vary: at low concentrations (below the crossover point indicated by the arrow) organism 1 will outgrow organism 2 (because of the higher affinity for the substrate), but at higher concentrations (above the crossover point) organism 2 will dominate because the maximum growth rate is higher than organism 1

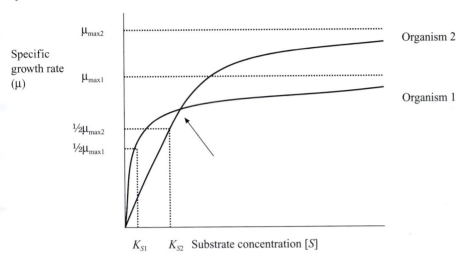

• **Figure 2.10** Temperature optima for bacterial growth. Organisms can be grouped according to the range of temperatures that they can grow in. Bacteria pathogenic to humans are predominantly mesophiles

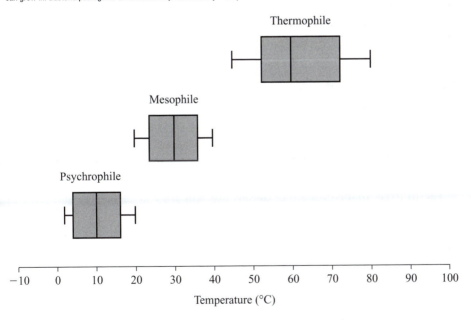

### ■ 2.4.2 GASEOUS ENVIRONMENT

Organisms can be grouped into three classes according to their tolerance of oxygen. It is somewhat surprising to discover that oxygen is a most toxic molecule. Oxygen is able to produce several highly reactive products that act as free radicals. Despite the fact that all mammals appear to thrive through air breathing, all life almost definitely arose from an anaerobic state. It was only as organisms acquired efficient mechanisms

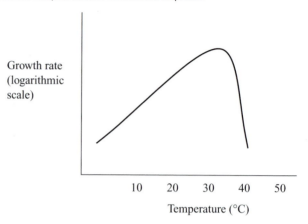

• **Figure 2.11** Growth rate of a mesophilic bacterium as a function of temperature

for detoxifying these reactive by-products of oxygen that the aerobic way of life started to proliferate.

### 2.4.2.1 Aerobes

These are organisms that grow in the presence of atmospheric concentrations of oxygen. Strict aerobes will not grow in the absence of oxygen.

### 2.4.2.2 Microaerophiles

Gaseous air contains approximately 20 per cent oxygen. Microaerophiles need reduced concentrations of oxygen (reduced oxygen tension) in order to grow and will not grow in air nor in the complete absence of oxygen. *Campylobacter* spp. will grow in 5–10 per cent oxygen.

### 2.4.2.3 Anaerobes

Obligate or strict anaerobes will not grow in the presence of very low concentrations of oxygen and many will also be killed. Different genera of anaerobic bacteria show a range of oxygen sensitivity. Anaerobes are unable to utilise oxygen for respiration and therefore will not grow in the presence of oxygen. Certain anaerobes may tolerate exposure to oxygen for a period, so a distinction has to be made between oxygen *killing* organisms and oxygen just *inhibiting* their growth.

The expression 'in the presence of oxygen' is inexact because the concentration will be a sliding scale rather than an all or nothing situation.

**Facultative anaerobes** are organisms that will grow in air but can also grow in anaerobic conditions. The considerable energy yields of aerobic over anaerobic metabolism are such that facultative anaerobes will preferentially grow in air and only switch to anaerobic metabolism when the oxygen is depleted. Facultative anaerobes will be the best equipped to deal with varying oxygen tensions, whereas the strict aerobes and strict anaerobes can be considered specialists that have adapted to particular gaseous environments.

These distinctions are useful and necessary in the cultivation and identification of organisms. In a natural environment, it is impossible to separate the oxygen concentration into three distinct levels. The reality is that numerous oxygen gradients must exist and organisms replicate at the preferred level (niche). The oxygen tension changes dramatically the closer you get towards a surface of the tooth or membrane surface of the intestinal tract. The small size of a bacterium will enable it to enclose itself appropriately in a gaseous mix of its choice. The immediate layers above the skin or mucous membrane will not be the same as the bulk environment.

There is a spectrum of tolerance to oxygen in anaerobic bacteria. This is reflected in the artificial, but useful, description of the range of oxygen tolerance:

extremely oxygen-sensitive anaerobes ———————— moderate ——— relatively oxygen-
tolerant anaerobes

Those anaerobes that are extremely sensitive to oxygen will be that much more difficult to culture and maintain in culture media than those that will tolerate exposure to oxygen. The exact mechanism(s) by which oxygen is damaging to anaerobic bacteria is not clear. Indeed, oxygen is toxic to prokaryotic and eukaryotic cells but they appear better equipped in limiting the damaging effects of oxygen. Oxygen is used in a limited number of cellular events, most importantly as the terminal electron acceptor in aerobic respiration (see above, pp. 67–71) which might reflect the potentially damaging properties. Oxygen is known to form highly **reactive free radicals (reactive oxygen species, ROS)** such as superoxide ions, hydroxyl radicals and hydrogen peroxide. Free radicals are defined by the possession of unpaired electrons in their chemical structure. Such a property makes them very reactive as they urgently seek to gain or lose electrons in order to reach a more stable configuration. This toxicity is harnessed to good use in eukaryotic cells where ROS are produced in the lysosome of neutrophils to attack and degrade the engulfed bacteria. Oxygen readily accepts electrons and thus acts as a potent oxidant. Hence oxygen, if fully reduced (complete addition of electrons and hydrogen), will yield water:

$$O_2 + 4H^+ + 4e^- \rightarrow 2H_2O$$

However, partial reduction of oxygen will result in formation of ROS:

$$O_2 + 2H^+ + 2e^- \rightarrow H_2O_2 \text{ (hydrogen peroxide)}$$

$$O_2 + e^- \rightarrow O_2 \cdot^- \text{ (superoxide radical)}$$

Free radicals, whilst highly reactive and hence damaging, are relatively short lived. The ability of organisms to protect themselves against these radicals may be one of the most important developments by which organisms were able to utilise oxygen. The production of the enzymes catalase and superoxide dismutase will help detoxify the radicals according to the following reactions:

$$O_2 \cdot^- + O_2 + 2H^+ \xrightarrow{\text{Superoxide dismutase}} H_2O_2$$

A further problem arises with the production of hydrogen peroxide because it can break down into another radical, the hydroxyl radical, as follows:

$$H_2O_2 + H^+ + e^- = H_2O + OH\cdot$$

It is crucial then that catalase and peroxidase enzymes act to catalyse conversion of hydrogen peroxide to water:

$$2H_2O_2 \xrightarrow{\text{catalase}} 2H_2O + O_2$$

It would be nice and tidy to be able to say that the relationship between the presence of superoxide dismutase and catalase and resistance to oxygen in bacteria is that straight-forward. Unfortunately it is not. Aerotolerance in anaerobic organisms has not always been shown to correlate with possession of catalase and superoxide dismutase. Detoxi-fying enzymes are likely to be significant in helping bacteria deal with oxygen, but the complete picture remains unclear as yet.

Anaerobes are abundant not least because oxygen is poorly soluble in water, thus permitting anaerobic and microaerophilic organisms to grow according to the oxygen tension of preference. Figure 2.12 shows how the oxygen tolerance of bacteria can be reflected in the areas of the broth culture that become turbid as the organism grows.

The concentrations of oxygen that develop in water are low. In distilled water at 25°C the concentration is less than 0.3 mM. By way of comparison the oxygen concen-tration (more commonly referred to as oxygen tension) in venous blood is 0.05 mM. The solubility of oxygen in water increases as the temperature decreases. Conversely, increasing the ionic strength of water with salts decreases solubility of oxygen. The consequence is that bacterial culture media, with the numerous salts, buffers and sub-strates that are added to encourage growth will severely limit the oxygen solubility.

The redox potential is a quantitative measure of the electron affinity (oxidising power) of a substance (Box 2.2). With oxygen being a strong oxidising agent (electron acceptor), as the concentration of dissolved oxygen increases, the redox potential ($E_h$) will be pulled towards a more positive value. The $E_h$ is thus a scale by which the tend-ency of a substance to donate or receive electrons relative to a standard reference (hydrogen) electrode is measured. The hydrogen electrode is used as the zero reference point. The readings obtained in millivolts will be negative if the substance is reduced, whereas positive values will be obtained from oxidising compounds.

Standard solutions of $O_2/H_2O$ have an $E_h$ of $\sim +820$ mV. Most anaerobes are inhibi-ted at $E_h$ values more positive than $-100$ mV. Extremely oxygen-sensitive anaerobes will not grow at $E_h$ values more positive than $-300$ mV. The measured redox potential

• **Figure 2.12** Growth of bacteria in broth according to their tolerance of oxygen. When cultured in fluid medium, the bacteria grow at the zone where the oxygen tension is optimal for the organism

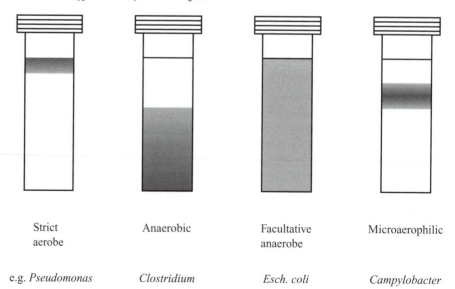

| Strict aerobe | Anaerobic | Facultative anaerobe | Microaerophilic |
|---|---|---|---|
| e.g. *Pseudomonas* | *Clostridium* | *Esch. coli* | *Campylobacter* |

---

### ■ BOX 2.2 OXIDATION AND REDUCTION

In general terms, catabolism is an exergonic process in which the substrate is oxidised. Whilst 'chemical oxidation' may mean addition of oxygen, in biological/biochemical terms, 'oxidation' refers to the removal of hydrogen or electrons. The compound that loses these reducing equivalents is oxidised and the acceptor is termed 'reduced' (addition of hydrogen or electrons). In other words, oxidised compounds lose electrons (or hydrogen ions) whereas reduced compounds gain electrons (or hydrogen ions). This process couples a reduced Donor (Donor$_{red}$) with an oxidised Acceptor (Acceptor$_{ox}$) as follows:

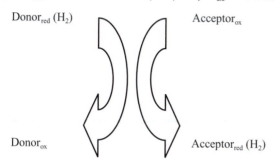

$$\text{Donor}_{red}\ (H_2) \qquad\qquad\qquad \text{Acceptor}_{ox}$$

$$\text{Donor}_{ox} \qquad\qquad\qquad \text{Acceptor}_{red}\ (H_2)$$

The oxidation will result in the transfer of two electrons ($2e^-$) and two protons ($H^+$) because

$$H_2 = 2H^+ + 2e^-$$

Thus oxidation is equivalent to losing two hydrogen atoms ($2 \times H = H_2$).

The compounds that undergo these complementary exchanges are called 'redox couples'. The readiness to transfer electrons varies between different redox couples. By measuring the tendency to undergo a redox reaction under standard conditions, it is possible to arrange the redox couples according to their redox potential. The movement of the reducing equivalents can be measured and compared to a standard redox couple – the hydrogen electrode. The accumulation of electrons is measured as a voltage (in millivolts) and given as $E_0$, a scale created in a similar manner to that of pH. When the conditions are set to have physiological relevance, i.e. pH 7 and at 37°C, then the symbol is altered thus: $E_0'$.

The reader is recommended to consult Wrigglesworth (1997) for a more detailed exploration of this important topic.

---

of a culture medium will broadly correlate with the ability of the medium to reduce oxygen, but the reading will be a sum of several redox couples within a complex (undefined) culture medium. As with oxygen toxicity and possession of catalase and SOD, the relationship between oxygen concentration in a solution or culture medium and the measured redox potential is not that straightforward. Oxygen toxicity may be due to increased (less negative) $E_h$ in the cytosol such that reduction of redox couples becomes impossible. For culturing anaerobic bacteria the aim is to lower the redox potential of the culture medium. This is achieved in the following ways.

• Removing oxygen from the gaseous environment

The cultures are placed in a jar or anaerobic cabinet and the air above the cultures is replaced with 80 per cent nitrogen, 10 per cent $CO_2$ and 10 per cent hydrogen. This can be achieved by removing the air with a vacuum pump and replacing it with the above gas mixture or using a chemical reaction to generate the appropriate conditions. The chemical reaction converts the oxygen into $CO_2$ and is catalysed by the addition of palladium pentoxide.

- Adding reducing agents

Certain chemicals have been used that are able to help maintain the $E_h$ at a desired level, in a manner comparable to pH buffers. They will attempt to set the $E_h$ of the medium close to their own $E_h$. Commonly-employed reducing agents include cysteine hydrochloride, sodium thioglycollate and dithiothreitol. To monitor the extent to which the cultures are being kept anaerobic, a colour dye can be added to the media. The indicator dye resazurin is colourless when reduced but turns pink when oxidised at pH 7.

### ■ 2.4.3 pH
Bacteria that live on and infect man grow best at pH 7 (neutral pH) and may be described as neutrophiles. Acidophiles prefer low pH, less than pH 6, whereas alkaliphiles grow at alkaline pH values over 8. Most organisms will tolerate a range of pH values that extend either side of their pH optimum as a bell-shaped curve (although the shape of the plot need not necessarily be symmetrical). The organisms will possess adaptive mechanisms with which to deal with the limits of tolerance, not least because the organisms themselves will force pH changes through their production of acids or bases as metabolic waste products.

### ■ 2.4.4 WATER ACTIVITY ($a_w$)
Whilst most weight of living matter is predominantly water, not all of this water is available for use. Agar plates that have dried out through sitting in a 36°C incubator for a week clearly have less water content than those that are freshly prepared. This can be seen in larger colony sizes in fresh culture plates. However, high concentrations of ions will also reduce the free available water content. The dissolved ions are kept in solution by surrounding themselves with a hydration shell of water. If the concentration of solute is high then less free water will be available for use by the bacterium.

The available free water, called **water activity** ($a_w$), that supports most bacteria must fall below values of 0.98, where pure water is 1.00. $a_w$ is defined as the ratio of the vapour pressure of the substance over that of pure water. The difference in water content between an apple and a peanut is easy to consider without resorting to measurement, but the $a_w$ may not be so easy to judge. For example, different food products vary quite considerably in the available water; for example, salami, 0.85 and dried cereals, 0.70. Low $a_w$ values will limit the number and extent of growth of microbes in the food and this is important in food storage considerations. Yeasts and moulds are generally able to grow in lower $a_w$ environments ($a_w$ of less than 0.8) than bacteria.

### ■ 2.5 MEASUREMENT OF BACTERIAL GROWTH
The answers to most questions asked by laboratory microbiologists involve counting the numbers of organisms in samples. Is there a sufficient number of organisms in this chicken meat to cause food poisoning? Is this organism killed by this antibiotic or can you culture a particular bacterium on culture media? Most experiments will involve

determining the growth or not of an organism. How is this done? To highlight the problems and limitations of the following methods, we can consider two samples of food that we need to examine for bacterial contamination. The first sample is a chocolate-coated biscuit, the second sample is a glass of milk. How easy is it to count bacteria in these samples?

### ■ 2.5.1 MICROSCOPIC COUNTING

The counting of bacterial numbers microscopically over time is the most direct means of demonstrating bacterial growth. This simple method involves placing a suspension of the organisms onto a counting chamber. The chamber has a grid etched onto the surface and the bacteria are observed directly through a microscope and counted. The number of bacteria is then calculated from the number observed within a set area (i.e. number of squares). The number in that area will represent the number in a volume of fluid that covers that area, hence the numbers are expressed as organisms per volume (e.g. $1.2 \times 10^5/l$). However, this method is unable to distinguish between live and dead cells (hence this is a total count, consisting of live and dead cells rather than a viable count which only counts living bacteria) and between small bits of dust, debris and marks on the face of the counting chamber. Furthermore, small bits of biscuit will be indistinguishable from bacteria. The identity of the bacteria will also remain unknown, simple morphology alone will tell you very little.

### ■ 2.5.2 COLONY COUNTS

Bacteria are easily observed when they grow as colonies on culture media. The assumption is made that one colony results from the growth of one bacterial cell. This is not necessarily the case as bacteria notoriously clump together. Nevertheless, a colony will only develop from viable bacteria; dead bacteria will not grow, divide and form colonies! Thus a plate count will tell you the total viable count. The original sample is usually diluted and a known volume is inoculated onto the surface of an agar plate. Knowing how much the sample has been diluted enables you to count the numbers on the plate and multiply this back by the dilution factor (see Figure 2.13). An additional advantage is that the purity of the culture can be assessed by comparing the colonial morphology of the colonies on the plates.

### ■ 2.5.3 BACTERIAL MASS

As discussed above, the measurement of bacterial growth need not necessarily just mean the counting of bacterial numbers. Several methods have been devised that use markers of the increase in bacterial mass. They can be grouped into measurements of

- **biomass** (dry weight, wet weight),
- **chemical constituents** (total protein, nitrogen, nucleic acid).

Such methods are indirect estimates of bacterial numbers and give no idea of the bacterial species involved or viability of the culture, but nevertheless, can provide a more rapid measure of bacterial mass than dilution counts. Obviously, the concentrations of nitrogen or protein of non-bacterial origin in the culture media needs to be corrected for (taking our example of milk renders this technique almost useless). Using such methods, the time taken for doubling of cell mass is termed 'doubling time' ($t_d$) rather than generation time as the latter term implies the increase in bacterial numbers rather than bacterial mass or product.

• **Figure 2.13** Estimating bacterial numbers by serial tenfold dilution steps. The test sample contains $10^4$ organisms/mL. 1 mL of the test is removed and diluted 1 in 10. The 1 in 10 dilution tube is then diluted similarly. This is continued along the row of tubes as necessary. To count how many organisms there are in each tube, 100 μL is removed and spread over the entire surface of a agar culture plate. After incubation overnight the number of colonies should reflect how many single bacterial cells there were in each tube. Because we know how many organisms there were in the test, we can predict how many colonies there will be on each plate. In the laboratory we will not know the number of organisms, but we can estimate it by working back. The first plate that yields 20–80 colonies is multiplied first by the dilution factor. Then the number is corrected to give colony forming units per mL (remember, you only put 100 μL onto the Petri dish). Thus: 10 colonies per 100 μL from the 1 in 100 dilution. This will be 10 × 10 (from 100 μL to 1 mL) × 100 (dilution factor) = $1 \times 10^4$ cfu/mL

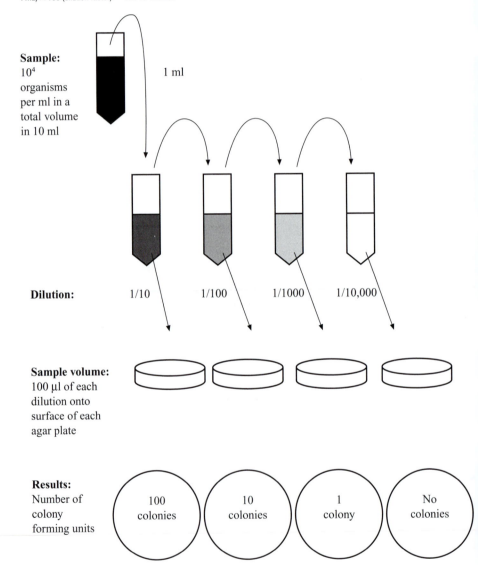

**Sample:**
$10^4$
organisms
per ml in a
total volume
in 10 ml

1 ml

**Dilution:**  1/10    1/100    1/1000    1/10,000

**Sample volume:**
100 μl of each
dilution onto
surface of each
agar plate

**Results:**
Number of
colony
forming units

| 100 colonies | 10 colonies | 1 colony | No colonies |

## ■ 2.5.4 LIGHT SCATTERING METHODS (TURBIDOMETRY, NEPHELOMETRY)

Bacteria in suspension will appear cloudy. Light passed into a suspension of bacteria will be deflected (scattered) in proportion to the number of bacteria present (Figure 2.14a). The ratio of the incident light ($I_0$) to that which emerges out the other side (transmitted light, I) can be converted into **absorbance** according to the following equation:

$$A = \log I_0/I$$

Absorbance is measured in a spectrophotometer. The shorter the wavelength of the light used, the greater the sensitivity will be. Unfortunately, short wavelengths are absorbed by cell components such as nucleic acid and proteins, and this will artificially

• **Figure 2.14** Optical density measurement by light scattering. (A) The light source is passed through the bacterial suspension. The deflected light is collected in the detectors. The extent to which the light is scattered is proportional to the numbers of organisms. (B) Relationship between absorbancy (A) and increasing bacterial mass or bacterial numbers measured at different wavelengths. The relationship should be linear (solid lines) but, in practice, the absorbance underestimates the high density of organisms (dashed lines)

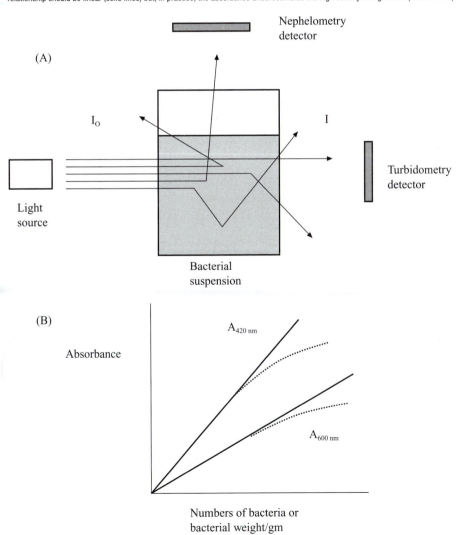

increase the absorbance reading, falsely overestimating the numbers of organisms in the sample. It is usual therefore to use wavelengths near 600 nm. At this wavelength with turbid suspensions of organisms the relationship between absorbance and bacterial numbers also deviates, but in the other direction (Figure 2.14b). Appropriate dilution is required to prevent underestimating the actual numbers of organisms. Older texts will mention the use of nephelometers. These machines operate on the same principle but capture the transmitted light at 90° from the incident light, rather than in line, as with a spectrophotometer. A newer variation on light scattering has been the use of **flow cytometry** (Figure 2.15). The method focuses a light source through a bacterial suspension flowing through a fine tube. The light source generates three parameters, a laser light source with a wavelength to measure fluorescence as well as light for measuring the light scatter in two directions (forward- and side-scatter). The fluorescence will be used to detect those organisms that have been stained with the particular fluorescent dye. The light scatter gives information on cell size and number and the fluorescence signal will enable particular properties of the cell to be measured (e.g. nuclear staining) or detect fluorescence-labelled organism out of a mixed population of bacteria.

### ■ 2.5.5 MOST PROBABLE NUMBER (MPN)

This is an old method, still in use today, in which samples of water or milk are taken and diluted in a suitable culture medium and, after incubation, observing for growth in the tubes. Two assumptions are made:

1. the organisms are randomly distributed in the sample being examined (water or milk), and
2. any viable organism in the test sample will grow (there are no inhibitory factors that may kill organisms during the process).

By making a series of dilutions from our sample, the pattern of growth can be referenced against a table of standard dilutions and the MPN is given for all the possible combinations.

• **Figure 2.15** Flow cytometry. The bacteria in suspension are passed along a tube through which a beam light source is focused. The scattered and incident light are collected in detectors 1 and 2 respectively, along with any fluorescent emissions from cells that have taken up a fluorescent stain (e.g. a nucleic acid stain such as propidium iodide). The use of fluorescent antibodies permits detection of specific organisms within mixed populations

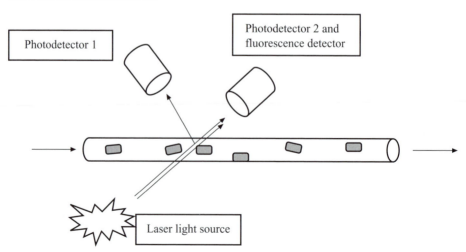

The statistics that underlie the MPN are explained through an understanding of the **Poisson distribution**. The Poisson distribution is a modification of statistics known as the binomial theorem. If it is assumed that a large number of samples have been examined but the likelihood of a positive result is small (e.g. a particular bacterium present in drinking water), then the Poisson distribution is appropriate to describe the pattern of results that are likely to occur. Poisson distribution is therefore sometimes referred to as the 'law of small probabilities'.

Poisson distributions are characterised by the following features:

- whole number observations (i.e. integers, not fractions),
- each object/number counted is a small fraction of the total number,
- successive observations/objects are independent of each other (occur randomly),
- the chances of the two outcomes do not vary from one observation to the next.

Note, however, that micro-organisms in nature do *not* exist in populations that can be described by a Poisson distribution. For example, consider the natural populations of micro-organisms, existing mostly as dense populations in biofilms.

Mathematically, the Poisson distribution is:

$$P = \frac{m^x \cdot e^{-m}}{x!}$$

where P is the probability of the sample containing one or more organisms;
$m$ is the *average* number of organisms per sample (being an average this cannot be measured directly);
$x$ is the actual number of organisms per sample; and
$x!$ is the factorial of $x$ (i.e. $n! = 1 \cdot 2 \cdot 3 \cdot 4 \ldots n$).

We have a suspension of 1000 bacteria in 10 ml. We set up four dilutions in tenfold steps ($10^{-1}$, $10^{-2}$, $10^{-3}$ and $10^{-4}$) and 1 ml volumes from each are placed into broth for incubation overnight. We can predict that the organisms will be distributed in the tubes as follows:

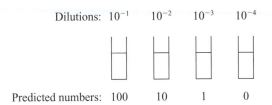

Dilutions: $10^{-1}$    $10^{-2}$    $10^{-3}$    $10^{-4}$

Predicted numbers: 100    10    1    0

The actual observed numbers will vary from the predicted because of the random and independent nature of the distribution of organisms in the original culture. In a dilute sample, the Poisson distribution can be used to estimate the probability that a tube will be sterile (will have no organisms present) and the probability that a tube has a mean number of organisms.

The probability of the sample being sterile will mean that $x$ is zero. Substituting into the Poisson distribution we obtain:

$$P(0) = e^{-m}$$

because $0! = 1$ and $m^0 = 1$.

If $P(0) = e^{-m}$ is the probability that nothing grows, then the probability that $x$ number of organisms will be present (and grow) will be the absolute certainty, i.e. $P = 1$, less the probability that nothing grows. This is represented as follows:

$$1 - P(0)$$

Or substituting from the equation above:

$$1 - e^{-m}$$

Now we can apply this to the Poisson distribution to state that the probability of more than one or more organisms in the sample will be:

$$P(\geq 1) = 1 - e^{-m}$$

This is useful when assessing how many organisms are present in the sample. The example given above used only one tube per dilution. In practice, the more replicate tubes used per dilution the better. If we used 10 tubes at each dilution and found that at $10^{-3}$ dilution we had five that were turbid after incubation, then $P = 0.5$ (i.e. 50 per cent).

We can rearrange the equation for $m$:

$$P = 1 - e^{-m}$$

$$P + e^{-m} = 1$$

$$e^{-m} = 1 - P$$

and take natural logs of both sides

$$-m = \ln(1 - P)$$

$$m = -\ln(1 - P)$$

In this format we can simply insert the proportion of tubes that grew as P. For example, if 50 per cent grew then $P = 0.5$, hence:

$$m = -\ln(1 - 0.5) = 0.69 \text{ organisms per sample}$$

**Table 2.3 The numbers of bacteria per tube ($m$) according to the most probable number derivation of Poisson distribution**

| Proportion of tubes turbid | $m$ |
| --- | --- |
| 20% | 0.22 orgs/sample |
| 40% | 0.51 orgs/sample |
| 60% | 0.91 orgs/sample |
| 80% | 1.6 orgs/sample |

Using this equation we can estimate the mean number of organisms in the sample by noting the proportion of tubes that are sterile. A number needs to be adjusted so as to account for the dilution factor. This is the principle of the most probable number (MPN) method.

The accuracy of the MPN method is dependent on the number of samples tested being large.

## ■ 2.6 CULTURING BACTERIA *IN VITRO*

When studying the structural features of a bacterial cell, it is possible to observe the structural components within a single bacterial cell using, for example, an electron microscope. To assess functional activity within bacteria we need to test a population of cells so as to generate a sufficient signal in our experiment (e.g. pH indicator change, colony formation on a plate, etc.). We are therefore testing the predominant characteristic of a population of bacterial cells.

Bacteria can be cultured using:

- liquid culture media (broth cultures),
- solidified agar plates, or
- culture media that contain both (biphasic).

Broth cultures usually permit the growth of small numbers of organisms where inoculation onto plates can fail. Broth cultures can be shaken so as to improve oxygen penetration and release the bacterial gas end-products. The biggest drawback of broth cultures, however, is that you cannot tell if the culture is mixed or not. In other words, how many bacterial species are present in the turbid suspension? You may have inoculated one organism but others could have been introduced inadvertently when the lid was opened.

Solid cultures offer a number of advantages over broth cultures. The colonies formed by the bacteria have particular patterns (colonial morphology) which can aid in differentiating between bacterial species. This gives a good idea of how many species were present in the sample. A widespread method of culturing bacteria is the batch culture. A small inoculum is introduced aseptically into a set volume of culture media and incubated appropriately. Under such conditions the typical growth curve is obtained (Figure 2.4b). Whilst convenient, batch culture has several problems. The organisms are in a continual state of change as they move from lag to exponential to stationary phase, and the organisms are not all in the same stage in their growth cycle. With time the culture becomes growth limiting through the depletion of a nutrient. Remember the loss of just one nutrient will limit the growth of the culture.

The alternatives to batch culture have considerable intellectual advantages. It is often desirable to study bacteria under steady state conditions in which the specific growth rate is kept constant during a prolonged period of exponential growth. In batch culture, exponential growth can be maintained by replenishing fresh media periodically. The pattern of growth obtained would, in theory, resemble the results shown in Figure 2.16. If the method was working perfectly, a saw-tooth pattern should be obtained with the slope of the curve identical throughout. The only parameter that changes is the concentration of organisms. In practice, such a method can be automated such that when the optical density (absorbance) of the culture reaches a certain limit then fresh culture medium is added. This method is the basis of the **turbidostat**. An alternative mechanism is found in the **chemostat** which is shown in Figure 2.17. As a **continuous culture** system, the chemostat will maintain the organism in exponential growth phase under constant conditions that can be easily altered to choice. Pre-warmed sterile

• **Figure 2.16** Maintenance of exponential growth by periodic addition of fresh culture media. Once the culture reaches exponential phase of growth, the serial dilutions can start. Every time the culture exceeds the threshold absorbance reading (dashed line), fresh culture media is added (arrows). Note that the rate of increase is constant, whereas the absolute numbers of organisms changes. The idea of continuous growth monitored by turbidity is the basis of the turbidostat

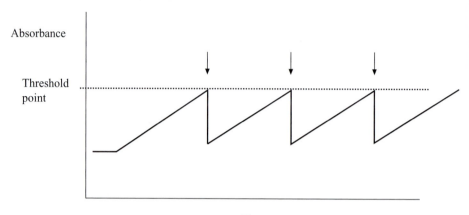

• **Figure 2.17** The chemostat. The volume of fresh medium added equals that collected at the overflow. The rate at which fresh medium dilutes the culture will also regulate the specific growth rate ($\mu$) because the additional nutrient is consumed almost as fast as it enters the culture (i.e. the nutrient is rate limiting). With a fixed rate of new medium the culture is in balanced growth but as the growth of the organism can only proceed as fast as the fresh medium is introduced, $\mu$ can be regulated by changing the flow rate.

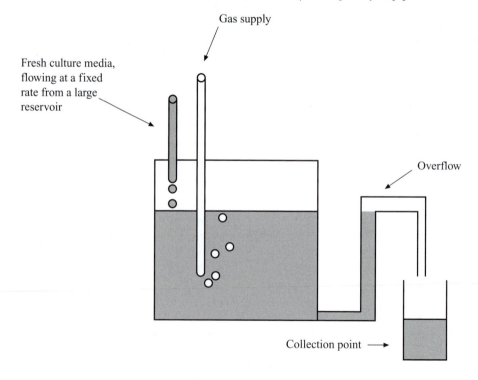

medium is pumped into the culture at a steady flow rate whilst the volume is maintained at a constant through the removal of the culture at the same rate as the inflow. When the rate of adding new medium (the dilution rate) is matched by the growth rate, the concentration of the organisms in the chemostat will remain constant (unlike the saw-tooth method!) as well as the growth rate.

In contrast to the degree of detail that has been worked out concerning bacterial metabolic pathways, the routine culture of bacteria in laboratories is often little short of mysticism. The use of culture media with defined ingredients is rarely necessary when cheaper, **undefined** or **semi-defined media** will suffice. The media used to culture organisms must be considered as highly artificial in comparison to the nutritional conditions encountered by organisms when growing in the natural environment, be that in the environment or on man. In addition to nutrients it is necessary to add **buffering** capacity. Phosphates and acetates will have buffering action to delay the pH of the culture moving outside the normal limits. Bacteria will tolerate some degree of acidity or alkalinity if in stationary phase of growth but during lag phase they are more susceptible.

For growing bacteria on solid agar, the addition of **gelling agents** are needed. The classic petri dish system employs **agar** to solidify the culture media. Agar is extracted from seaweed and is a polysaccharide.

Continuous culture methods are controlled by different methods:

- the rate of input of a limiting nutrient: the chemostat,
- the density of the culture: the turbidostat.

The properties of agar that make it so useful are that it melts at 84°C but forms a clear gel at around 40°C and this process can be repeated several times. Agar is not pure but a mixture of the polysaccharides, agarose and agaropectin; the ratio of these two determines the rigidity and viscosity of the gel. Being of natural origin, it contains traces of ions (e.g. $SO_4^{2-}$) and metabolites. Agar has a net anionic charge and thus binds cations such as $Mg^{2+}$ and $Ca^{2+}$. These details should highlight that agar is not an inert substance.

### ■ 2.6.1 DIFFERENTIAL MEDIA

To aid the identification of specific organisms on culture media, it is common practice to add particular compounds and reagents to culture media so that the colonies on the plate appear distinct from the other contaminating organisms. A very widely used culture medium called 'MacConkey agar' is used to isolate *Escherichia coli* from urine when investigating patients with urinary tract infections. Another organism that can also cause urinary tract infections is *Proteus mirabilis*. MacConkey agar is used because it contains lactose as the sole source of carbon along with a pH indicator. As *Escherichia coli* ferments lactose, it causes the pH to drop and the pH indicator turns purple, whereas *Proteus mirabilis* does not ferment lactose and does not alter the pH. The colonies of *Escherichia coli* thus appear purple and those of *Proteus mirabilis* remain the colour of the unchanged medium.

### ■ 2.6.2 SELECTIVE MEDIA

Such media add selective agents to the culture to attempt to prevent the growth of the other organism present in the sample under investigation. Some of the selective agents commonly used are:

- **dyes** such as crystal violet at low concentration tend to inhibit Gram positive organisms and thus permit the growth of Gram negative bacteria.
- **bile salts** are a useful selective agent. Gram negative bacteria are able to tolerate higher concentrations of bile salts than Gram positive bacteria, and within Gram negative bacteria differences exist in their tolerance.
- **salt**, 7–10 per cent (w/v) NaCl is used as a selective agent in the culture of *Staph. aureus*. Many other bacteria are unable to tolerate this salinity, unlike *Staph. aureus*. Ironically, the use of salt as a preservative therefore is compromised when dealing with enterotoxin producing *Staph. aureus* and goes some way to explaining the high proportion of food poisoning outbreaks from salty foods such as salami.

### ■ 2.6.3 ENRICHMENT CULTURE

There are a number of situations when you will need to try to grow a particular bacterium from a patient but the numbers of organisms are low, both within the patient and within the sample. Enrichment cultures aim to first culture the sample so that sufficient numbers of the organism are reached that, when it is subsequently cultured onto the usual culture plates, you will observe the targeted organism in large enough numbers to see it easily.

## ■ 2.7 FUELLING GROWTH: BACTERIAL METABOLISM

For bacteria to grow, they need appropriate environmental conditions such as correct temperature, gaseous requirements (physicochemical parameters), in addition to a suitable energy source. The nutrients are mostly dissolved in water and absorbed as aqueous solutions. The enormous diversity of situations from which bacteria can be recovered is mirrored by an appropriate diversity of substrates and the metabolic pathways to deal with them. It is not the intention of this chapter to review this metabolic diversity but to concentrate on the broad metabolic patterns of a bacterium of which most strains exist without any harm in the intestine but others are causes of serious illness, i.e. *Escherichia coli*.

The metabolic pathways of both prokaryotes and eukaryotic organisms are built around a common set of metabolic pathways that supply a pool of building block, precursor compounds along with the use of ATP as an energy carrier. Bacteria have existed for a much longer time than eukaryotes with the appearance of eukaryotic cells occurring relatively recently in a geological timescale. Hence the metabolism of eukaryotes is usually seen as developmental modifications of a prokaryote system. This chapter will take a broad look at bacterial metabolism and again concentrate on the organisms that are associated with humans. Details of biochemical pathways and energy production are not covered as these are amply dealt with in other texts.

Catabolism: oxidative in character.
Anabolism: reductive in character.

Reducing equivalents produced in catabolism are then utilised in anabolic pathways.

The sum of all the cell's chemical reactions are collectively termed 'metabolism' but can be more instructively split into **catabolism** (the reactions yielding energy) and **anabolism** (those that require energy). There are three key ingredients to cellular metabolism: carbon, reducing equivalents and energy, and the careful manipulation of these three ingredients permits the cell to drive biosynthetic (anabolic) reactions and other energetic events by the energy harnessed from catabolism.

### ■ 2.7.1 CARBON

Bacteria are known to utilise a wide range of organic compounds as their source of carbon, including carbohydrates, proteins and lipids of differing complexity. This variety of substrates is in marked contrast to the small number of metabolites that are produced as a result of their catabolism. Bacteria do not possess distinct metabolic pathways for each substrate but will convert the substrate into a form that is utilised by a small number of common, or central pathways.

### ■ 2.7.2 REDUCING EQUIVALENTS

This somewhat unfriendly term simply describes the protons ($H^+$) and electrons that are transferred as a hydrogen atom between compounds during catabolism and anabolism. The transfer of reducing equivalents will occur via specialised compounds, of which the most important are nicotinamide adenine dinucleotide ($NAD^+$) or the phosphorylated form $NADP^+$ (see Box 2.3). The $NAD^+$ and $NADP^+$ shuttle protons and electrons from substrates during catabolism and transfer them to anabolic pathways and feed hydrogen atoms into the electron transport chain.

---

### ■ BOX 2.3 REDUCING EQUIVALENTS AND THE ELECTRON CARRIER NAD⁺/NADH + H⁺

$NAD^+$ is the oxidised form and NADH is the reduced form. Reducing equivalents will reduce $NAD^+$ according to the following reaction:

$$NAD^+ + 2H \rightarrow NADH + H^+$$

The two hydrogen atoms can be written as the component parts to emphasise the reducing equivalents donated by the appropriate molecule in the metabolic pathway:

$$NAD^+ + 2H^+ + 2e^- \rightarrow NADH + H^+$$

More accurately, only the hydride ion ($H^-$: a proton and 2 electrons) is transferred and thereby neutralising the charge. This is why the spare proton is written separately for the reduced NADH:

$$NAD^+ + H^- + H^+ \rightarrow NADH + H^+$$

---

### ■ 2.7.3 ENERGY

Catabolism and anabolism are coupled so as to utilise the energy released from the former to fuel the latter, but this is not the only requirement for energy. The cell still needs to function. Some of the energy-requiring activities of bacterial cells are given in Box 2.4. These energy-requiring activities have been termed 'work'. The use of a separate term from energy serves to highlight two points: bacteria need mechanisms to convert energy into functional work and the energy obtained from catabolism that is directly coupled to anabolic reactions will not be available for other energy-requiring processes. These activities need to obtain energy from a more general source.

The breakdown of the substrates (catabolism) is coupled to the anabolic synthesis of necessary products required for functioning and growth of the bacterial cell. The coupling typically occurs through the cyclical degeneration and regeneration of ATP:ADP and $NAD^+$ and NADH. The substrates for catabolism vary tremendously, depending on the bacteria under investigation, but for those that infect humans the sources are organic compounds, and are given as carbohydrates, proteins and fats. As a generalisation, catabolism is mostly oxidative whereas biosynthesis is largely reductive, i.e. involves the incorporation of reducing equivalents donated from NADPH and utilises ATP.

According to the sources of energy, reducing equivalents and carbon can be used to group bacteria into types. The terms used to describe the different nutritional sources are shown in Table 2.4. The terms are constructed from whether the energy source is

---

### ■ BOX 2.4 CELLULAR ACTIVITIES THAT UTILISE ENERGY THAT ARE NOT SOLELY DERIVED FROM CATABOLIC PATHWAYS

- **Cell growth**: the manufacture of new cell constituents (including genetic material), cell division, spore formation.
- **Membrane function**: maintenance of the ionic gradients across the cytoplasmic membrane. Solute transport via different energy-requiring transport systems.
- **Movement**: flagellar rotation.

**Table 2.4 Nutritional classification of bacteria**

| Energy source: | light: | phototrophic |
| | chemical: | chemotrophic |
| Electron source: | inorganic compounds: | lithotrophic |
| | organic compounds: | organotrophic |
| Carbon source: | $CO_2$: | autotrophic |
| | organic: | heterotrophic |

supplied by light (**phototroph**) or from the energy released from chemical bonds (**chemotroph**). Depending on whether the chemical bonds are in inorganic or organic compounds the organisms are termed **lithotrophs** or **organotrophs** respectively.

The bacteria that infect humans appear to obtain their energy as chemoorganotrophs, that is, they use organic compounds for energy and carbon sources. The organic compounds utilised vary between carbohydrate, protein (as amino acids), fatty acids, purines and pyrimidines but many bacteria associated with humans favour glucose as a principal fuel with amino acids as a second most common substrate. The breakdown of the larger organic compounds will be used to generate metabolic energy in a suitable form, such as ATP, and provide a number of compounds that are used as basic building precursors in most bacteria. This common pool of high-energy intermediates (including pyruvate, acetyl coenzyme A, phosphoenol pyruvate) are created from oxidative processes (loss of hydrogen or electrons) and coupled to a reduction reaction in which the oxidant ($NAD^+$ or $NADP^+$) receives hydrogen or electrons (to become NADH or NADPH). These reduced compounds are correspondingly used to fuel anabolic reactions. The overall scheme for heterotrophs is depicted in Figure 2.18.

The increased size of eukaryotic cells has permitted the specialisation of certain

• **Figure 2.18** Generalised plan of heterotrophic microbial metabolism. The carbohydrates, proteins and lipids will all be catabolised into key intermediates so that they can enter the fermentation or respiration pathway reactions in order to yield ATP. The exact patterns will differ between organisms but the central pathways utilise many of the same enzymes, reflecting the unity of biochemistry across prokaryotes and eukaryotes

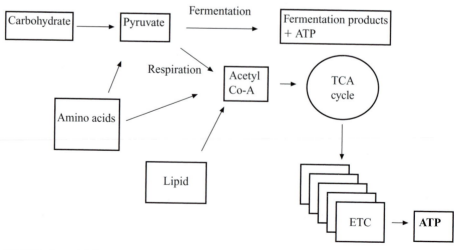

ETC: electron transport chain
TCA cycle: tricarboxylic acid cycle
ATP: adenosine triphosphate

**Table 2.5** Cellular location of the various key metabolic activities in prokaryotes and eukaryotes

|  | Cytosol | Inner cytoplasmic membrane |
|---|---|---|
| Prokaryotes | glycolysis<br>fermentation<br>TCA | electron transport chain |

|  | Cytosol | Mitochondria |
|---|---|---|
| Eukaryotes | glycolysis<br>fermentation | electron transport chain<br>TCA |

energetic pathways to particular organelles; for example, respiratory electron pathways are centralised to mitochondria in eukaryotic cells. The differences in location of the various metabolic processes are given in Table 2.5.

Bacteria require energy to fuel the maintenance of the cell and, if conditions are favourable, growth and division. This energy is taken from two processes:

- chemical energy via molecules such as ATP obtained by **substrate level phosphorylation (SLP)**, and
- electrochemical energy via **electron transport chain phosphorylation (ETP)**.

Let us first look at substrate level phosphorylation. Molecules such as ATP possess phosphate groups which are readily released from the parent compound and donated to a recipient molecule, a process known as 'phosphorylation'. ATP is the most widely known example of a so-called 'high energy' or 'coupling' compound. Adenine triphosphate will cycle between ATP and ADP such that the available (free) energy is released when the phosphate group is lost (an **exergonic** reaction) and free energy is regained when the phosphate group is added to ADP to form ATP (an **endergonic** reaction). Catabolism is the process used to energise the production of ATP from ADP, whereas anabolism will utilise this energy by coupling the phosphorylation of ATP to drive other biosynthetic reactions. As most people are aware, glucose is a readily utilisable metabolic fuel. Carbohydrates are the principal energetic substrates for many bacteria but, despite the large number of different compounds that bacteria can utilise as energy sources, the catabolic pathways employed by the bacteria converge into a few key intermediates. In *Escherichia coli* there are a few favoured catabolic pathways that result in the production of these energy intermediates. The important pathways are:

- the Embden–Meyerhof–Parnas pathway (which is usually referred to as glycolysis),
- the pentose phosphate pathway (PPP), and
- the Entner–Doudoroff pathway (ED).

The details will be found in most biochemical textbooks and will not be given here.

The second mechanism for ATP production mentioned above is electron transport chain phosphorylation (ETP) (Figure 2.19). The energy for regenerating ATP via ETP is collected from the ion gradients across the cytoplasmic membrane. Movement of hydrogen ions ($H^+$) across the cytoplasmic membrane (from high concentrations outside the cell to low concentrations inside) occurs via the membrane-located enzyme

• **Figure 2.19** The generation of ATP is coupled to the electron transport chain. The resultant accumulation of protons (H⁺) on the external face generates a proton gradient across the cytoplasmic membrane. This proton motive force is used to generate ATP as protons flow through the ATP synthetase towards the cytosol

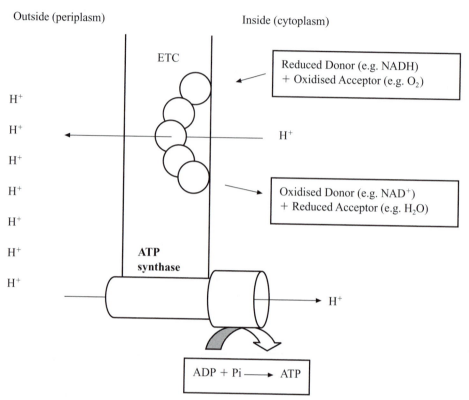

**ATP synthase.** The extrusion of protons (H⁺) occurs through the electron chain, a series of electron carriers that move electrons via a series of redox reactions. ETP therefore occurs through the combined action of the electron transport chain and the ATP synthase. The former creates the potential energy supply in the form of a proton motive force (pmf) and the latter then harnesses that potential energy to drive ATP synthesis and other energy-requiring activities (see margin note). Put another way, the ETP converts energy from coupled redox reactions ($\Delta E$ in millivolts) to a proton motive force ($\Delta p$, also in millivolts) to free energy ($\Delta G$) which will be used as energy to drive phosphorylation.

The ETP consists of an independent (i.e. not part of the catabolic pathway in question) series of compounds (proteins and lipids) which are redox couples. These compounds can be grouped as hydrogen carriers (NAD⁺/NADH) and quinones (ubiquinones and menaquinones, highly lipid soluble compounds and electron carriers: cytochromes (cytochromes a,b,c,d and o).

The electrons pass along the ETP from redox couples with low redox potentials (very negative values, e.g. NAD⁺/NADH = − 300 mV) along to quinones, then to the cytochromes and finally to the terminal electron acceptor that has highest redox potential (very positive values such O₂/H₂O = +820 mV). If the terminal electron acceptor

**Margin notes:**

Cellular activities that utilise energy that are not solely derived from catabolic pathways

**Cell growth:** the manufacture of new cell constituents (including genetic material), cell division, spore formation.

**Membrane function:** maintenance of the ionic gradients across the cytoplasmic membrane. Solute transport via different energy-requiring transport systems.

**Movement:** flagellar rotation.

in the series is oxygen, the process is called **aerobic respiration**. If the terminal acceptor is anything other than oxygen, e.g. $NO_3$, $SO_4$ or fumarate, the process is called **anaerobic respiration**. Effectively, only bacteria exhibit anaerobic respiration, as very few eukaryotes have been shown to utilise anaerobic respiration.

Bacteria with an electron transport chain system are able to convert reducing equivalents into coupling compounds such as ATP.

The exact composition of the electron chain varies depending on which transport chain components they possess (cytochromes, quinones, etc.) and the terminal electron acceptor that is used. Bacteria can alter the sequence of the ETP according to the conditions they find themselves in. In this way, an organism has a pool of available ETP carrier molecules depending on the metabolic circumstances. Because of the greater yield of ATP with aerobic respiration over that of anaerobic respiration, in facultative anaerobes the presence of oxygen preferentially diverts metabolism towards aerobic pathways over the anaerobic routes. Only when oxygen concentrations are exhausted do the anaerobic pathways take over. In *Esch. coli*, for example, the preferred route will end with oxygen, but under anaerobic conditions it will use nitrate instead or, in the absence of utilisable nitrate, it will use fumarate.

## ■ 2.7.4 FERMENTATION

Rather than a vague term for brewing alcohol, fermentation defines a particular catabolic process. Fermentation is an oxygen-independent catabolic pathway in which an organic substrate is used to generate ATP via substrate level phosphorylation. In comparison with respiration, fermentation does not use an external oxidising agent but instead the end product itself is used as the terminal electron acceptor. In other words, fermentation reactions couple oxidation and reduction reactions in the pathway such that the reducing equivalents are passed from the oxidised substrate to the major end product of that pathway. For example, fermentation of glucose to lactic acid (via the Embden–Meyerhof–Parnas pathway) results in the formation of pyruvate. Pyruvate is then reduced to lactate and NADH is reoxidised back to $NAD^+$. The net gain from the whole pathway is 2 moles of ATP from substrate level phosphorylation and the organism is left with 2 moles of lactate to discard. The oxidation of glucose to pyruvate generated the ATP and NADH, and the step from pyruvate to lactate simply reoxidises the NADH to $NAD^+$. Hence, the net number of reducing equivalents within the pathway has not changed but simply cycled through the electron carrier $NAD^+/NADH$. The end product is excreted by the organism because it can no longer use it to generate further ATP. Other organisms may well be able to utilise the end product and in this way a mixed population of fermenting organisms will reduce compounds to end products such as $CO_2$ and hydrogen gas.

The substrates that can be fermented are not restricted to carbohydrates. Anaerobes can ferment alcohols, purines and organic acids. The fermentation pathways are usually named after the major end product. These end products are useful in classifying bacteria (Figure 2.20).

Most fermenters (typically, anaerobic bacteria) utilise substrate level phosphorylation because they do not possess a respiratory chain with which to undergo electron-transport reactions as a means of creating ATP. Whilst fermentation yields significantly less ATP than aerobic respiration, anaerobes will exploit the anaerobic environment that has been vacated by organisms that have adopted an aerobic existence. This presumably justifies the persistence in employing comparatively inefficient anaerobic fermentation.

• **Figure 2.20** Fermentation products (via EMP pathway). Pyruvate is the key intermediate in fermentation of carbohydrates via the EMP pathway. The cyclical oxidation and reduction of NAD$^+$ supplies the reducing equivalents. The incomplete oxidation of pyruvate will yield different products depending on whether the organism has the appropriate enzymes. The end products therefore can characterise different groups of organisms

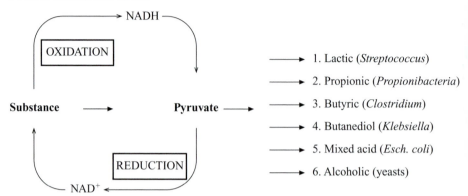

1. Lactic (*Streptococcus*)
2. Propionic (*Propionibacteria*)
3. Butyric (*Clostridium*)
4. Butanediol (*Klebsiella*)
5. Mixed acid (*Esch. coli*)
6. Alcoholic (yeasts)

## ■ 2.8 REGULATION OF METABOLISM

To maintain a sensible grip on the multitude of biochemical pathways that are present within bacteria, a series of regulatory controls are needed. The points at which the various regulatory mechanisms work varies. Likewise, the physical locations of the regulatory steps differ, as do the targets for regulation.

Certain enzymes are continuously synthesised (**constitutive enzymes**) unless instructed to stop, whereas other enzymes are only produced when specifically requested (**inducible enzymes**). Details of the various biochemical and genetic regulatory mechanisms can be found in appropriate biochemical texts. Regulation of metabolism can be grouped according to site of action or in more functional properties. Taking the former idea, the physical sites of regulation can be separated into three locations.

### ■ 2.8.1 THE GENOME: SYNTHESIS OF THE APPROPRIATE ENZYMES

A key step in regulating enzyme activity is to control the rate of transcription of the enzymes. In this way, only enzymes that are required are manufactured. This is more economical than destroying enzymes that have already been manufactured. Figure 2.21 shows in broad terms how the regulation can act to up regulate (enzyme induction) or down regulate (enzyme repression) enzyme synthesis. The *lac* operon is one example of regulation by negative control of a series of catabolic enzymes (Figure 2.22). The *lac* operon codes for three enzymes involved in the utilisation of lactose. The *lac* operon transcribes all three enzymes as one polycistronic mRNA and this mRNA is under the regulatory control of a single set of regulatory sites (promotor, operator and termination regions).

---

**■ BOX 2.5 THE ADVANTAGES OF THE OPERON SYSTEM**

- Control is exerted at the level of transcription and not translation.
- The grouping of all the related genes in close proximity under a single promotor permits the single polycistronic mRNA and is therefore very efficient.
- The single control point enables instant reaction to environmental signals.

---

• **Figure 2.21** Enzyme induction and repression. The end products in both (A) and (B) act by inducing or repressing the synthesis of the enzymes in a pathway. Enzyme induction (A) usually applies to catabolic pathways. The pathway is only required if the corresponding substrate is present. With enzyme repression (B), the pathway is usually anabolic and the enzymes are therefore required as long as the organism is growing. Thus, only when the end product is in excess are the enzymes repressed

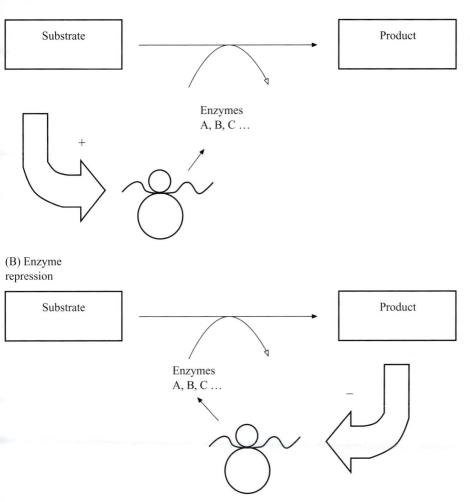

(A) Enzyme induction

Substrate

Product

Enzymes
A, B, C …

+

(B) Enzyme repression

Substrate

Product

Enzymes
A, B, C …

−

■ **2.8.2 THE CELL MEMBRANE: UPTAKE OF THE SUBSTRATE AT THE CELL MEMBRANE**

The proteins that regulate transport of nutrients can be up or down regulated. The control of these proteins occurs mostly at the level of transcription.

■ **2.8.3 THE CYTOSOL: THE ENZYME ACTIVITY**

Those regulatory events that act on the gene product rather then the gene itself are called **post-translational** control mechanisms. Figure 2.23 illustrates a feedback (end-product) inhibition mechanism. The term 'inhibition' rather than 'repression' is used to indicate that the regulation is post-translational, i.e. directly at the enzyme itself. The mechanisms involve alteration of the catalytic efficiency of existing enzymes

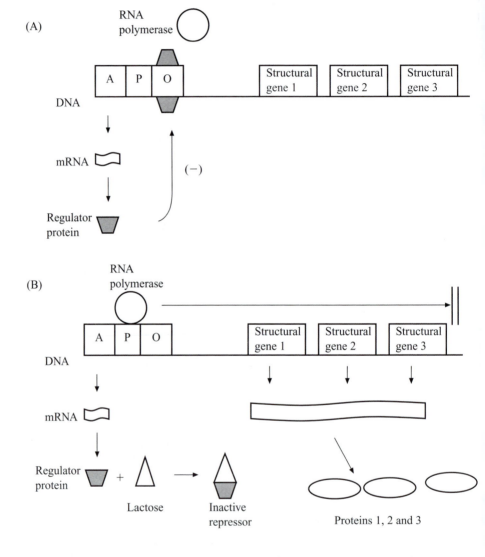

• **Figure 2.22** The lac operon. (A) When lactose is absent the constitutively produced repressor binds to the operator region (O) and transcription by RNA polymerase is largely prevented. However, in (B) when lactose is present it needs to be metabolised and transcription occurs. Lactose binds to the repressor and prevents it from binding to the operator site. This permits RNA polymerase to proceed from the promotor site (P). In this operon lactose is the effector or inducer molecule

by allosteric modulation. This will directly modify the efficiency of existing enzymes. A comparison of feedback inhibition and repression is given in Table 2.6.

The numbers of genes that are regulated in their expression can vary from single enzymes through to collections of genes related to a single function (operons) or multiple operons (global control). When changes in the concentration of one particular substrate may change but everything else remains reasonably constant, the organism will simply modify the appropriate operon. A regulatory control of more than one operon is carried out by a **regulon**. When the environment changes dramatically (the organism falls from the nutrient-rich skin of man into the rainwater on the floor) then there is a need to completely reorganise the metabolic pathways. Regulons illustrate how bacteria can regulate related functions with great efficiency but also give a sense of how cellular

• **Figure 2.23** Feedback (end-product) inhibition of enzymes. Enzyme pathways can be regulated by feedback inhibition where the end product inhibits the first enzyme in the pathway for its manufacture. The direct inhibitory effect on the enzyme will cause immediate effect on the synthesis of the substrate

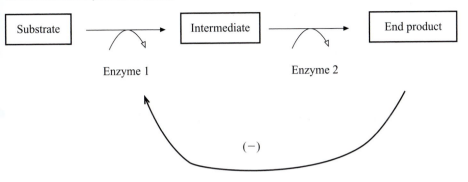

**Table 2.6 Comparison of feedback inhibition and repression**

| Feedback inhibition | Repression |
| --- | --- |
| Inhibition of enzyme activity | Inhibition of enzyme synthesis |
| Site of inhibition: | Site of inhibition: |
| the enzyme (protein) | transcription (DNA) |
| Usually only the first enzyme in a pathway is inhibited | All enzymes in the operon inhibited |
| Time course: rapid | Time course: slow |
| Effect: fine regulation | Effect: course regulation |

function is integrated at several layers. Higher level or global regulation are in turn regulated by **modulons** and other higher level multi-gene systems. Control of larger areas of metabolism provides a quick response to dramatic changes in conditions. The global regulatory networks are not solely concerned with metabolic events. Other stimuli that activate global networks induce the stress response system. The stimuli encompasses such events as heat shock (a sudden shift to a higher temperature), cold shock, osmotic shock (sudden shift to a high or low osmolarity) or sporulation. All of which require a range of different gene products to help cope with the environmental insults. Mostly the response induces synthesis of DNA repair and oxidation-protecting enzymes.

## ■ 2.9 GROWTH *IN VIVO*

If a single bacterium continued to divide unhindered for 8 hours, over $10^9$ organisms would accumulate. Given the vast numbers of organisms that are present even in the human gut, unlimited exponential growth would bury the planet under an enormous mass of bacteria. Clearly then, there are limits to such unrestrained proliferation. The constraining factors will include the quantity and quality of the nutrients and physico-chemical parameters such as water activity and temperature. In addition to nutrient restrictions, other lethal events can occur such that even if bacterial numbers are increasing they may be matched by rates of cell death; for example, toxic compounds secreted by other organisms or UV irradiation from sunlight. There are several factors that help limit the proliferation of bacteria within the human intestine. They include: mammalian inhibitory products (lysozyme, secretory IgA), bile salts, gastric acid and other bacterial inhibitory proteins (bacteriocins).

With the advent of the sequencing of entire bacterial genomes, it has become apparent that many bacteria do not possess all the enzymes in common metabolic (usually catabolic) pathways such as the TCA cycle. One possible explanation is that bacteria obtain key intermediates from other members of the microbial population. The term **syntrophism** is used to describe the relationship between two organisms that both have a metabolic dependence on the products from the other organism. Syntrophism is represented diagrammatically in Figure 2.24. Syntrophism is a way of reducing the total genetic load on an individual species because fewer enzymes are needed by each organism. It may also reduce competition between organisms in mixed communities.

### ■ 2.9.1 FEAST THEN FAMINE

Bacteria will experience periods of feast and periods of famine. The range of nutrients (carbon, nitrogen trace elements, etc.) may be inadequate or present for only short periods of time. Equally, key nutrients may be present in rate-limiting concentrations, in which case, the bacterial mean generation time will be considerably longer *in vivo* than that obtained *in vitro*.

For bacteria that reside as normal flora in man, they will reside in stable niches such as the gut or skin. It would seem likely that such organisms are going to encounter relatively steady supplies of nutrients in contrast to organisms that spread from one host to another via the environment. *Vibrio cholerae*, for example, is not a normal constituent of the human intestinal flora. Human cholera infections arise through the ingestion of *Vibrio cholerae* in faecal-contaminated water or food. The nutrient content of river water will differ both in type and concentration from that in the intestinal tract. Similarly the shedding of *Staphylococcus aureus* from the skin epithelia onto the floor of a

• **Figure 2.24** Syntrophism. Only organism 1 can catabolise compound A to produce compound B, but only organism 2 can catabolise compound B. Both organisms can catabolise compound C onwards but are dependent on one of the two intermediates (compounds B and C)

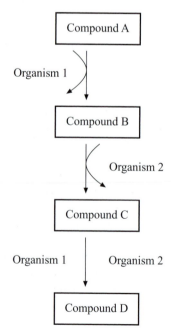

hospital ward before being accidentally transferred back onto the hands of a person will all represent nutrient-rich and nutrient-poor environments.

### ■ 2.9.2 IRON LIMITATION

A more specific example of nutrient limitation is that of iron. Most bacteria have an absolute requirement for iron, hence it is growth-limiting. In aerobic conditions, iron is present as $Fe^{3+}$ which is poorly soluble. This contrasts with anaerobic conditions where reduced iron ($Fe^{2+}$) is freely soluble. Concentrations of available iron in aerobic mammalian tissues are therefore correspondingly low. Complexed with host iron-binding proteins such as lactoferrin and transferrin, the available (free) iron in the body is approximately $10^{-18}$ M. Bacteria therefore give particular attention to mechanisms of extracting iron from the host in order to grow. Two mechanisms by which bacteria attempt to obtain sufficient concentrations of iron are given, one sophisticated, the other a little more destructive.

#### 2.9.2.1 Production of iron chelators

The iron scavenging compounds produced by bacteria (**siderophores**) vary in their ability to pull iron ($Fe^{3+}$) from host-binding proteins. To be effective siderophores need to have greater iron-binding affinity than the host compounds. Indeed, this is the case, with iron chelators of *Escherichia coli* having affinities greater than 10-fold compared to human transferrin. The siderophores are excreted by the bacterium into the environment where they bind iron and are then transported back into the bacterial cytosol across the cell membranes via specific transporter systems. Depending on the chemical type of siderophore, the iron is released either by hydrolysis of the siderophore or it is reduced to $Fe^{2+}$, a much more soluble form that does not complex with siderophores. The changes in solubility of reduced iron mean that anaerobic bacteria will not suffer from the problems of iron starvation like aerobes. Synthesis of siderophores is under negative control; that is, the synthesis is normally repressed until levels of iron in the bacterium drop below a critical point and then synthesis commences (i.e. is de-repressed).

Production of diphtheria toxin by *Corynebacterium diphtheriae* is regulated by the available free iron concentration. It is proposed that the diphtheria toxin lyses eukaryotic cells as a means of releasing iron.

#### 2.9.2.2 Production of haemolytic toxins

By disrupting eukaryotic cells, particularly red blood cells, haemolytic toxins will release iron into the environment.

### ■ SUMMARY

Bacteria divide by binary fission and the numbers will increase exponentially unless external constraints limit the increase (nutrient limitation, build up of toxic products). The rate of increase in bacterial mass or numbers can be measured and provide useful markers of bacterial physiology. In batch culture, bacteria exhibit a typical growth curve with various phases of growth described; however, this pattern of growth is unlikely to occur so readily in the natural environments. Three mechanisms are employed to generate energy: aerobic respiration, anaerobic respiration and fermentation. Knowledge of the metabolic requirements and other parameters such as temperature and gaseous conditions is essential for growing bacteria in culture media.

Evolutionary pressures mean that those organisms best adapted to the prevailing conditions proliferate. Competition for nutrients will select out those organisms that are best adapted to the prevailing conditions. The consequences for the less well adapted will be starvation.

## FURTHER READING

Dawes, I.W. and Sutherland, I.W. (1992) *Microbial Physiology*, 2nd edition, Blackwell Scientific Publications, Oxford, UK.

Lengeler, J.W., Drews, G. and Schlegel, H.G. (1999) *Biology of the Prokaryotes*, Georg Thieme Verlag, Stuttgart, Germany.

Neidhardt, F.C., Ingraham, J.L. and Schaechter, M. (1990) *Physiology of the Bacterial Cell*, Sinauer Associates, Sunderland, Massachusetts, USA.

Phoenix, D. (1997) *Introductory Mathematics for the Life Sciences*, Taylor & Francis, London, UK.

Stanier, R.Y., Ingraham, J.L., Wheelis, M.L. and Painter, P.R. (1986) *General Microbiology*, 5th edition, Macmillan, Hampshire, UK.

Sutton, R., Rockett, B. and Swindells, P. (2000) *Chemistry for the Life Sciences*, Taylor & Francis, London, UK.

(Further details of redox reactions and all chemical principles relevant to microbial physiology are explained here.)

White, D. (2000) *The Physiology and Biochemistry of Prokaryotes*, 2nd edition, Oxford University Press, Oxford, UK.

Wrigglesworth, J. (1997) *Energy and Life*, Taylor & Francis, London, UK.

(This book explains redox reactions and the variations in metabolic pathways and respiratory chains that can occur in both eukaryotes and prokaryotes.)

## REFERENCES

Hamilton, W.A. (1988) Microbial energetics and metabolism, in Lynch, J.M. and Hobbie, J.E. (eds) *Micro-organisms in Action: Concepts and Applications in Microbial Ecology*, Blackwell Scientific Publications, Oxford, UK.

Morris, J.G. (1990) The metabolism, growth and death of bacteria, in Parker, M.T. and Collier, L.H. (eds) *Topley and Wilson's Principles of Bacteriology, Virology and Immunity*, 8th edition, Edward Arnold, London, UK.

## REVIEW QUESTIONS

*Question 2.1*    What is the difference between the lag phase and the stationary phase in the typical growth curve obtained in batch culture?

*Question 2.2*    How does growth rate influence competition between bacteria?

*Question 2.3*    Is the pure culture a good model for studies of bacterial growth?

*Question 2.4*    What is the difference between 'total' and 'viable' count in bacterial culture?

*Question 2.5*    Is it better to let incubators run slightly below or above the optimum temperature when culturing bacteria? Why?

*Question 2.6*    What differences would you expect to see in the growth curves between two strains of *Escherichia coli* in broth culture inoculated at i) low density and ii) high density?

## ■ 3.1 VIRUSES, BACTERIOPHAGES, VIROIDS

The word 'virus' has origins in the words 'venom' and 'poison'. From early times, certain diseases such as rabies or smallpox were easily recognised as lethal to humans. The microscopic agents could pass through filters that retained bacteria and were therefore called filterable agents. Viruses are successful parasites; they infect all the main types of living organisms from animals, plants and insects to fungi and bacteria. Viruses that infect bacteria are called **bacteriophages**.

Viruses are **obligate intracellular parasites**. They have an absolute requirement on the host cell for manufacture of new virus components. Viruses do not multiply by binary fission but instead are assembled from component parts (nucleic acid and protein). This can only occur inside the host cell. Hence, viruses are inert outside of their hosts. With no need for a basal metabolic activity to retain viability, viruses have no need to carry ribosomes or other organelles or metabolic pathways. They simply are particles that can be copied and built by the cells that they infect.

With no intrinsic metabolic activity, viruses have lost any need for diffusion of gases and solutes. The limits on minimum size seen in bacterial cells are therefore removed. Viruses that infect man range between 20–200 nm in diameter, with most around 50–10 nm. At 200 nm, poxviruses (cowpox, smallpox) are large enough to be resolved by light microscopes and are comparable in size to the smaller bacteria. Viruses can be smaller than 10 nm but reduction in size will be limited by the length of nucleic acid necessary for replication and survival in the host.

In the process of becoming entirely parasitic, certain viruses have lost the minimum amount of genetic information such that they are unable to replicate inside the host cell unless aided by another virus. Such viruses are called **defective viruses**. Delta agent is a defective virus that infects humans and can only replicate in cells that have also become infected with hepatitis B virus. Certain viruses have lost so much nucleic acid that they are simply infectious strands of RNA. Called **viroids**, these agents lack even a protective protein coat surrounding the RNA strand yet are able to infect plants, transmitted following mechanical abrasion of the plant surface. To date viroids have not been implicated in human disease. In contrast to viroids, **prions** are replicating proteins involved in bovine spongieform encephalopathy (BSE) in cattle and the related degenerative brain disorder, kuru. Viroids and prions remind us not to state that viruses are the smallest infectious agents known.

Whether viruses are alive is not a particularly useful debate, hence only the important characteristics will be mentioned. Viruses resemble living organisms in that they have a genome and are able to replicate and evolve. Indeed, they have biological properties such as particular host ranges, routes of transmission and tissue tropism. Viruses, however, do not create or store free energy in compounds such as ATP and have no intrinsic metabolic activity outside their host cells (unlike spores or seeds) and are therefore not alive, at least for part of their existence. These comparisons place viruses in between chemicals and true living organisms. Attempting to include the important features, a definition of a virus is thus:

> A microscopic organism that invades and only reproduces inside living cells. Viruses possess one type of nucleic acid, are unable to replicate by binary fission but are assembled and do not undertake independent energetic metabolism.

### ■ 3.2 VIRAL STRUCTURE

Figure 3.1 shows the two basic structural types of virus that infect man. The variety of viral shapes would be much greater if animal, plant and bacterial viruses were included and so the term 'typical' virus is best avoided. At the simplest level a virus has just nucleic acid with a protein coat and some viruses also enveloping the nucleocapsid within a phospholipid envelope. Viruses are traditionally described as:

- **helical**,
- **icosahedral**, or
- **complex**.

Figure 3.1 shows the arrangement of the protein subunits (called **capsomers**) that form a suitable structure (**capsid**) to act as a protective coat for the nucleic acid against physical and chemical stresses. The shape of the structures formed are either helical or

• **Figure 3.1** Virus composition. (A) Helical capsids. The nucleic acid is contained within a cylinder made from proteins arranged in a helical stack. The protein shell is called a 'capsid'. (B) Icosahedral capsid. The nucleic acid is held within an icosahedral-shaped capsid. (C) and (D) The capsid of either form may be surrounded (enveloped) by a phospholipid membrane (the envelope)

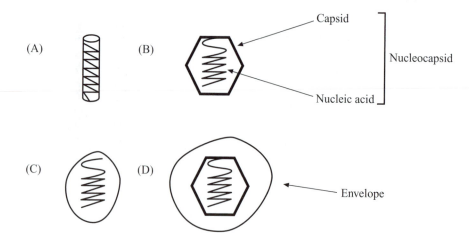

---

■ **BOX 3.1 HELIX OR SPIRAL?**

A helix is corkscrew shaped, with the curling motion rising up as well as round the central axis. A spiral will simply curl round a central axis but remain in the same plane. Wrapping tape around a pencil will form a spiral when viewed from the top of the pencil as it increases the total diameter with each circle completed.

---

icosahedral. The complex viruses may be a combination of the two or simply difficult to classify according to the above ideas.

### ■ 3.2.1 HELICAL VIRUSES

Nucleic acid will inherently form a helical shape (Box 3.1). The attachment of multiple protein capsomers will then form a helix as they follow the shape of the nucleic acid. The combination of capsid enclosing a nucleic acid is called a **nucleocapsid**. Viruses of this structure are also called **filamentous** or **rod-shaped** viruses (Figure 3.2). Helical symmetry will run along the axis of the helix. It is seen in viruses that possess this coiled, spiral-shaped nucleocapsid, common in human RNA viruses. A certain degree of flexibility is found in a helical arrangement of the capsomers that helps minimise the risk of the virus breaking in two. Helical viruses can regularly be observed in a curved shape. The length of filamentous viruses varies according to the length of the nucleic acid. The simplicity of using repeated subunits in a helix must explain why a large number of viruses adopt this configuration. Helical viruses that infect man are also covered in an additional lipid bilayer envelope. This envelope is added during replication of the virus, either as it passes through the host nuclear membrane or the cytoplasmic membrane.

### ■ 3.2.2 ICOSAHEDRAL VIRUSES

The majority of human and animal viruses are icosahedral. An icosahedron is a regular polyhedron (Box 3.2). Viral icosahedra are constructed from 20 equilateral triangles and

● **Figure 3.2** Helical virus shape. (A) The helical shape of the protein capsomers surrounding the helical strand of nucleic acid. This arrangement can be simplified diagrammatically as (B) where the nucleic acid is shown inside the capsid

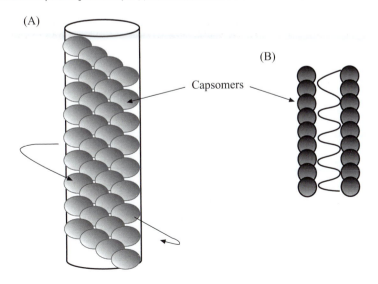

(A)

(B)

Capsomers

---

■ **BOX 3.2 POLYHEDRONS**

A polyhedron is a structure built up from repeated planar shapes. There are five regular polyhedrons, all of which can be placed within a sphere:

- tetrahedron: 4 triangular faces,
- cube: 6 square faces,
- octahedron: 8 triangular faces,
- dodecahedron: 12 five-sided faces,
- icosahedron: 20 triangular faces.

Of the five shapes, viruses utilise the icosahedron as it most closely approaches a sphere, thereby providing greatest strength.

---

have 20 flat surfaces, 12 corners (apexes) and 30 edges (see Figure 3.3). An icosahedron has three lines of symmetry: five-, three- and two-fold (Figure 3.3). The icosahedron is built to house and protect the viral nucleic acid. It could be built from a few large proteins or from many smaller proteins. Viruses employ the latter option because smaller proteins needs fewer bases and therefore less genome. Architects and engineers will tell you that an icosahedron is a highly robust shape that can be built from asymmetric sub-

• **Figure 3.3** The structure and symmetry of an icosahedron. The icosahedron can be arranged to show three axes of symmetry: 2-fold, 3-fold and 5-fold, and these are indicated as dotted lines

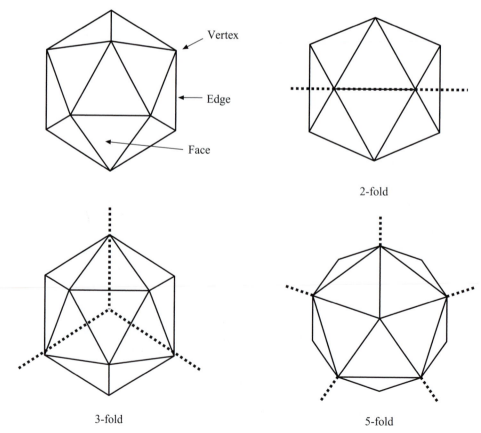

units but arranged to give a symmetrical result. Rather than form perfect spheres, viruses manufacture the icosahedron because it takes fewer components (capsomers) to form the more rigid icosahedron structure than a sphere. The protein capsomers that make up the equilateral triangle faces are not symmetrical themselves; they also do not possess straight edges. These asymmetric protein capsomers therefore arrange themselves like a jigsaw puzzle to form a symmetrical face in the shape of an equilateral triangle (Figure 3.4). Because the chemical bonding between the capsomers is identical, you obtain the greatest possible strength. By using very few proteins to build the capsomers, the virus maintains optimal economy in terms of genetic space (nucleic acid sequence) for the virus that needs to keep the nucleic acid as small as possible. The icosahedral viruses are also described as isometric, polyhedral or spherical viruses. Isometric (of equal measure) refers to the 'cubic' symmetrical pattern of the capsid.

### ■ 3.2.3 COMPLEX

Some viruses with no overall symmetry are nothing less than combinations of icosahedra and helices. For example, certain bacteriophages are composed of icosahedral heads attached to filamentous shafts yielding shapes that resemble the lunar space craft of childhood science fiction (Figure 3.5).

● **Figure 3.4** Assembly of an icosahedron. The diagram illustrates how the assembly of three asymmetric proteins can be grouped together to form a robust, symmetrical structure within which the viral nucleic acid is protected

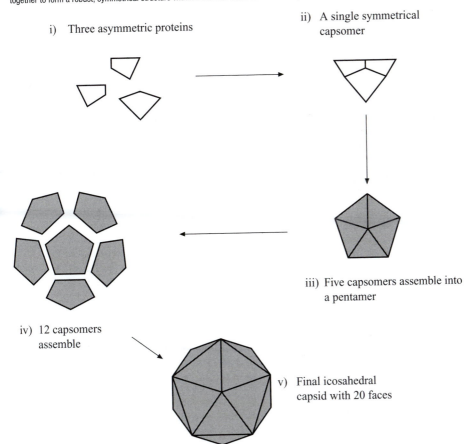

i) Three asymmetric proteins

ii) A single symmetrical capsomer

iii) Five capsomers assemble into a pentamer

iv) 12 capsomers assemble

v) Final icosahedral capsid with 20 faces

• **Figure 3.5** Complex morphology of a bacteriophage. The complex morphology of the bacteriophage can be viewed as an icosahedron (head) coupled to a cylinder (sheath). After landing on the bacterial surface, to infect the bacterium, the phage needs to inject the nucleic acid through the cell wall and cytoplasmic membrane. The phage acts like a syringe and contracts the sheath in order to inject the nucleic acid

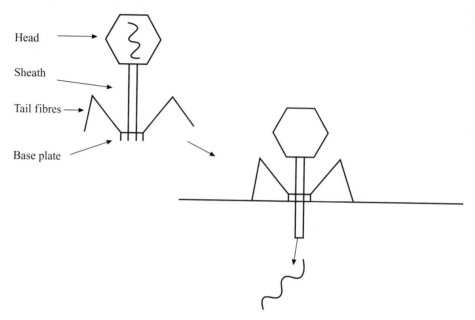

Head

Sheath

Tail fibres

Base plate

### ■ 3.2.4 NUCLEIC ACID

One of the most unusual features of viruses is the presence of only one type of nucleic acid. Whereas other organisms possess both DNA and RNA, viruses have only one nucleic acid type, DNA or RNA, never both. The implications of having only RNA as the sole genetic code are seen in the later section on replication strategies. Figure 3.6 illustrates the possible arrangements in viral nucleic acid. It follows that the following arrangements can be generated:

- single strand DNA,
- single strand RNA,
- double strand DNA,
- double strand RNA.

• **Figure 3.6** Viral nucleic acid

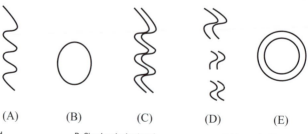

(A)    (B)    (C)    (D)    (E)

A: Linear single strand      B: Circular single strand      C: Linear, double strand
D: Linear segmented      E: Circular double stranded
Note that the hydrogen bonding that holds the double stranded forms together is not shown

In general, the single strand DNA format is the most unusual, the other three arrangements are all well represented in viruses that infect humans. The nucleic acid need not be kept in single strand; influenza virus, for example, has a **segmented** genome consisting of eight strands of single strand RNA.

With the limitations on size, viruses are usually haploid and code for relatively few proteins. The largest viruses (pox viruses) have only 200 genes (approximately 250 kilobases), whereas small viruses code for less than 10 genes! Viruses have several tricks for generating the greatest number of products out of the smallest number of coding sequences:

- **overlapping reading frames.** The start point for translation of two different gene products overlap. This avoids coding for nucleotides that are not translated.
- **genes within genes.** Small sequences may exist within a larger gene sequence. Problems of which gene is translated are avoided because the smaller gene is encoded in a different reading frame.
- **RNA splicing.** The mRNA that is transcribed can be modified by removing a small section of the mRNA and joining the two pieces back together again (splicing). Splicing occurs normally with eukaryotic RNA in the 'spliceosome', the virus simply exploits the host process as usual. Splicing results in a second protein formed from the original mRNA, both of which will have been read from the same promotor (Figure 3.7). Alternatively, by removing a section of the (pre-)mRNA, splicing can change the reading frame (and generate a second protein) or remove a stop codon and permit the translation of a larger protein (in the same reading frame).

• **Figure 3.7** Three mechanisms by which a single viral RNA transcript can be manipulated to yield more than one gene product

i) Original mRNA

ii) Shortened mRNA

iii) Section removed to generate two mRNAs

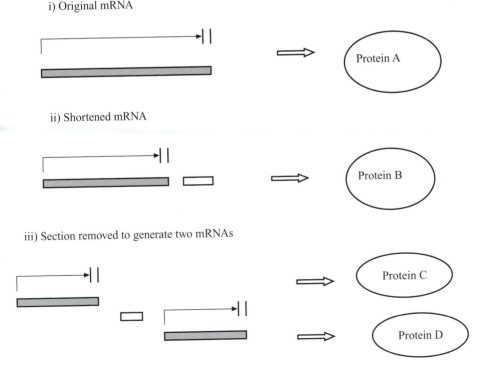

### ■ 3.2.5 ENVELOPE

Many helical and icosahedral viruses are contained within a phospholipid membrane envelope. The envelope is not manufactured by the virus (this would be expensive in terms of gene space and host cell synthesis) but simply taken from the host cell nuclear or plasma membrane when the virus leaves the cell. This means that the envelope is not pure phospholipid but contains proteins and glycoproteins that were present in the host envelope.

The possession of an envelope has quite a marked effect on the properties of a virus. The envelope is mostly phospholipid and therefore relatively delicate. The phospholipid envelope is highly susceptible to drying.

Most helical viruses that infect humans are enveloped. Naked helical viruses of humans are not known. Enveloped viruses can exit from a cell via budding (see below, p. 95), whereas non-enveloped viruses often need to lyse the cell in order to escape. The reason is not clear but perhaps this strategy is too destructive as a mechanism of maintaining itself as an infectious agent in humans.

### ■ 3.3 VIRAL CLASSIFICATION

As happened with many bacteria, older classification systems named the virus according to the disease it caused; e.g. poliomyelitis and poliomyelitis virus, or viruses isolated from adenoid tissues gave the term 'adenovirus'. This system is fine for species level but is unable to group viruses into higher taxonomic groups such as Genus, Family and so on. With increased understanding of viral structure, viruses could be classified according to the nature of their components:

* type of nucleic acid,
* size (diameter) of capsid (and number of capsomers),
* capsid symmetry (icosahedral, helical, complex or asymmetric),
* presence or absence of envelope,
* site of assembly within host cell (nucleus or cytoplasm).

Considerable sense was introduced by the introduction of the **Baltimore system** where the variations in nucleic acid content provided the basis for grouping viruses:

Group I: dsDNA viruses
Group II: ssDNA viruses
Group III: dsRNA viruses
Group IV: (+) strand RNA viruses
Group V: (−) strand RNA viruses
Group VI: RNA reverse-transcribing viruses

This system reflects the replication strategy leading to the production of mRNA. It may be criticised for concentrating on the replication strategy and ignoring all else; nevertheless, the explanatory power of the Baltimore system is demonstrated in the discussion below (pp. 89–93).

### ■ 3.4 VIRAL REPLICATION

Viruses must replicate within the cells of the infected host. Viruses infect not just humans but animals, plants, fungi and bacteria. We will consider the replication of viruses that infect humans. Bacteriophages, for example, replicate in a manner that is distinct from mammalian virus replication. For example, note that bacteriophages inject only the nucleic acid into the host bacterium whereas the entire virus article is internalised in mammalian cells.

The virus-infected cell should be seen as a controlled hijacking of the normal cell by an intruder. The replication strategy will depend on the type of genome, and the release of the virus will determine the pattern of infection within the host.

A number of differences are apparent when comparing viral replication with bacterial replication (Figure 3.8). A virus replicates itself from scratch, starting from the transcription of its nucleic acid, whereas a new bacterial cell is derived from a pre-existing bacterial cell as it doubles its cellular components and divides, a process not driven from nucleic acid transcription. This is partly reflected in the time taken to form new viruses and bacteria. A bacterium dividing in culture can be a relatively short event (20 minutes for *Esch. coli*), whereas the replication of a virus takes roughly 8 hours to complete. Whilst the bacterial cell doublings can be estimated during exponential bacterial

• **Figure 3.8** Comparison of growth curves for bacteria and viruses when grown *in vitro*. Note that the viral growth curve plots the numbers of free virus, i.e. virus that is in the culture medium but not those that are replicating within the cells being used to grow them

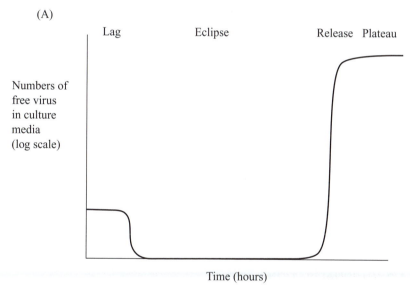

(A)

Lag    Eclipse    Release Plateau

Numbers of
free virus
in culture
media
(log scale)

Time (hours)

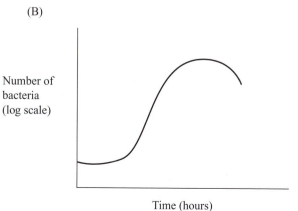

(B)

Number of
bacteria
(log scale)

Time (hours)

growth, numbers of viruses produced from a single virus within one host cell cannot be predicted. The integration of the host cell machinery with that of the virus makes the targets for antiviral compounds all the more difficult. Without wishing to underestimate bacterial replication, the process of binary fission of bacteria after duplication of all the bacterial contents appears more straightforward than the replication and assembly of a virus particle.

The replication of the virus can be divided into five stages, reflecting the general sequence of events:

- **attachment**,
- **internalisation**,
- **transcription of the genome**,
- **virus assembly**,
- **virus release**.

### ■ 3.4.1 ATTACHMENT

A specific structure on the virus particle will be the ligand that binds to the appropriate receptor on the membrane of the host cell. The viral-binding component will bind specifically to this host cell membrane receptor. The viral ligand will be a protein or glycoprotein but the host cell receptor is not restricted to a protein base. The target-binding site on the host cell will have its 'normal' function, independent of the viral binding. The virus exploits this receptor to use it as a means for establishing attachment. The nature of the receptor molecules varies but many of the receptors used by viruses are involved in immunological signalling (complement or immunoglobulin-binding molecules). The viruses will need to bind to a receptor on the host cell plasma membrane, but this is not necessarily the trigger for entry into the host cell cytoplasm. The ligand on the virus needs to be exposed. In non-enveloped viruses this presents no real problem; this ligand is present on the capsid surface, but in enveloped viruses the phospholipid membrane will require the ligand to be expressed within the envelope or project out from the envelope on stalks (e.g. influenza).

Antibodies directed against the viral-binding ligand will protect against infection and are called 'neutralising antibodies'. The binding ligands are thus desirable targets for vaccination.

### ■ 3.4.2 INTERNALISATION

Once bound, viruses need to cross the host cell plasma membrane in order to reach the replication machinery in the nucleus and cytoplasm. This problem is overcome through several mechanisms, partly dependent on whether the virus is enveloped or not. Two mechanisms have been identified:

- **membrane fusion**, where enveloped viruses fuse their membrane with that of the host cell and the capsid is then exposed to the cytosol (Figure 3.9), and
- **viropexis**, where the virus is taken up by an **endosome** (Figure 3.9).

Endosomes are vesicles (sacs) of membrane that are pinched from the plasma membrane and fused with lysosomes. Ligands that bind to certain receptors trigger the process of **endocytosis** in which the virus bound to the plasma membrane are engulfed into the cytoplasm. Within the endosome the virus still needs to escape through the plasma membrane wall of the endosome in order to reach the host cell

• **Figure 3.9** Viral entry. (A–C) show the process of viral fusion with host membrane. (D–G) illustrates the process termed 'viropexis'. The virus enters the host through endosomes (illustrated on the membrane by the vertical lines). The viral nucleic acid reaches the cytoplasm only after translocation across the endosomal membrane (G). Note that in viropexis, translocation is dependent on the exposure to acidic pH in the endosome, whereas fusion is not pH dependent. The solid black line represents the viral nucleic acid

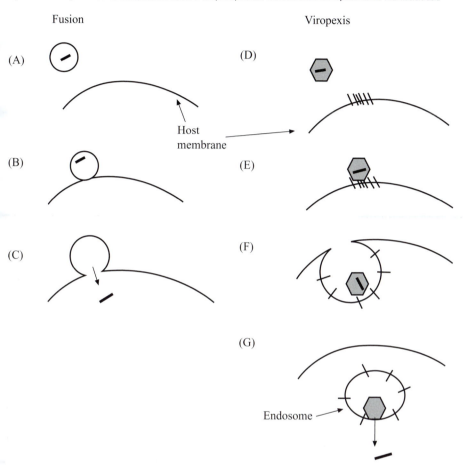

replication machinery in the cytosol. With influenza viruses this occurs because the acidic conditions in the endosome induce a viral protein to undergo a conformational change in its 3D structure so as to form a transmembrane pore, through which the viral components enter the cytosol.

Non-enveloped viruses may directly pass through the membrane, but the mechanism by which this occurs remains mysterious.

## ■ 3.5 VIRAL NUCLEIC ACID DETERMINES REPLICATION STRATEGY

### 3.5.1 TRANSCRIPTION OF GENOME

Once in the cytosol, the capsid needs to be uncoated so as to allow transcription of the nucleic acid. The strategy for replication will depend on the type of nucleic acid carried by the virus. DNA viruses will be treated in the same way as DNA of the host but because RNA viruses come in two types (positive and negative strand viruses), they are treated separately. Positive strand RNA act as mRNA in that they can be translated into proteins directly at the ribosome. Negative strand RNA, on the other hand, is not a functional RNA but the complementary copy (as seen with complementary strands of DNA).

During the process of uncoating, the capsid will be broken down so as to expose the viral genome for transcription and translation. This period of time will mean that there is no intact virus left within the cell following entry into the cell. If the cells are broken open, no intact, and therefore, infectious virus will be released. Hence, this period is called the 'eclipse phase'. The equivalent point on the growth curve for bacteria will be the lag phase; only bacteria are still viable during the lag phase (see Figure 3.8).

### ■ 3.5.2 DNA VIRUSES (FIGURE 3.10)

The viral DNA will be transcribed into mRNA by the normal processes of the host cell. This involves synthesising mRNA from the viral DNA strand using the host enzymes. The host enzyme is called **double strand DNA-dependent RNA polymerase II**

• **Figure 3.10** Replication strategy of dsDNA viruses. Upon entry into the host cell, the viral DNA is replicated by viral DNA polymerase transcribed within the nucleus by host transcriptases. The viral DNA also codes for capsid proteins and the new virion is assembled in the nucleus ready for release

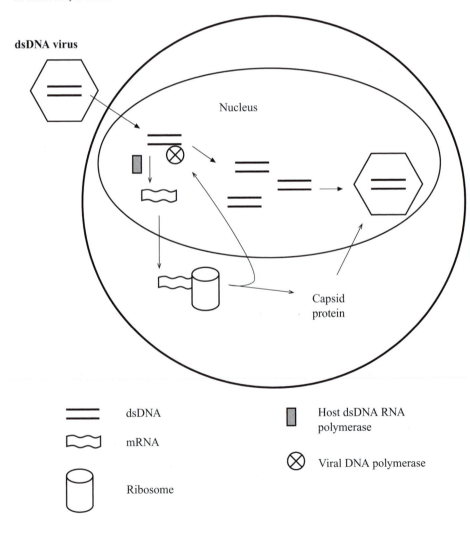

**dsDNA virus**

Nucleus

Capsid protein

| | | |
|---|---|---|
| ≡ | dsDNA | ▮   Host dsDNA RNA polymerase |
| 〰 | mRNA | ⊗   Viral DNA polymerase |
| ⬭ | Ribosome | |

(ds DNA-RNA polymerase). The term describes the target (double strand DNA) and the function: to catalyse the production of multiple strands of mRNA (RNA polymerase).

The mRNA codes for two types of protein: i) structural proteins to build the capsid and ii) DNA polymerase, the enzyme that catalyses the production of multiple copies of the original viral DNA strand, to be packaged into the new virion.

DNA viruses need to be transcribed within the nucleus in order to mirror the normal host cell processes. The structural proteins are transported into the nucleus and the new virus is assembled.

Single strand DNA viruses follow the same strategy but need to first synthesise the complementary strand of their DNA using the host cell polymerase so that transcription can proceed.

### ■ 3.5.3 POSITIVE STRAND RNA VIRUSES (FIGURE 3.11)

Once the genome is uncoated, it can be read by the host ribosome to produce two key products (as seen in DNA strand viruses): viral capsid proteins and an enzyme with which to replicate multiple copies of the original (+) strand RNA. The enzyme for the latter process is a **single stranded RNA-dependent RNA polymerase**. Remember

• **Figure 3.11** Replication strategy of positive strand RNA virus. The viral nucleic acid can be translated at the ribosome directly. The products are viral RNA polymerase to catalyse synthesis of new + strand RNA as well as capsid proteins for new virus

**Positive strand RNA virus**

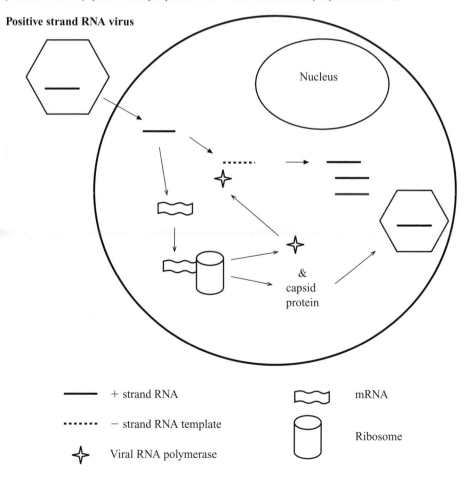

the mammalian cell cannot replicate RNA from RNA, therefore the virus has to code for the necessary RNA polymerase. In order to replicate a single strand of RNA, it is first necessary to manufacture the complementary (−) strand. This is then used as a template from which the (+) RNA strand is copied and packaged into the new virus particles.

These processes all occur in the cytoplasm of the cell. The viral (+) RNA therefore has all the properties/functions that enable itself to be replicated (i.e. the mRNA itself is infectious once in the cytoplasm).

### ■ 3.5.4 NEGATIVE STRAND RNA VIRUSES (FIGURE 3.12)

These viruses have a genome that will not be read directly as mRNA and are thus termed negative strand. To act as mRNA, the original strand will need to be converted into the complementary, functional strand. As mentioned above, mammalian cells are

• **Figure 3.12** Replication strategy for negative strand RNA virus. The viral RNA cannot be translated at the ribosome until the viral encoded RNA polymerase catalyses the synthesis of complementary positive strand RNA

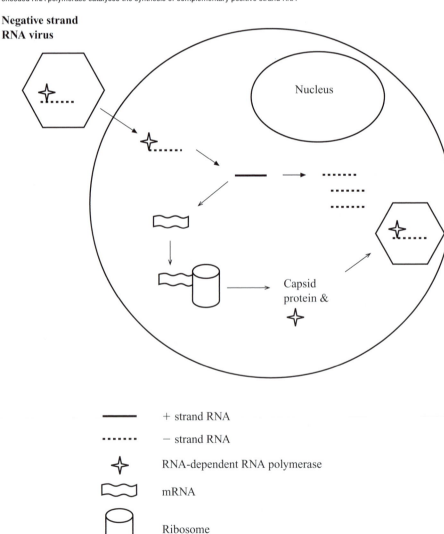

Negative strand
RNA virus

+ strand RNA

− strand RNA

RNA-dependent RNA polymerase

mRNA

Ribosome

unable to do this (because they generate mRNA from DNA) so the virus has to provide the appropriate polymerase so that the viral genome can be read. This RNA-dependent RNA polymerase is contained within the infecting virus particle so that upon uncoating it can immediately copy the (−) strand into (+) strand and only then can translation of the viral genome begin. In (+) RNA viruses, the RNA-dependent RNA polymerase is produced directly from translation but with (−) strand viruses it needs to be packaged ready in the capsid.

Now that the (+) strand RNA can be translated, capsid proteins are manufactured along with RNA-dependent RNA polymerase so as to create multiple copies of new (−) strand RNA from the (+) strand. Both the new copies of (−) strand RNA and the polymerase are packaged as new virus.

With RNA viruses these events occur in the cytoplasm. The necessity of having the RNA polymerase in the capsid means that the (−) strand RNA alone is incapable of self-replication and thus not infectious, even if injected into the cytoplasm.

### ■ 3.5.5 RETROVIRUSES (FIGURE 3.13)

The impact of the human immunodeficiency virus (HIV) on world health is a sobering reminder that the importance of understanding viral replication strategies cannot be underestimated. HIV is a retrovirus, so named because of the remarkable mechanism it uses to replicate its RNA genome. The central dogma of molecular biology had been that RNA is made from DNA and that this information flow was one way: DNA could not be created from an RNA template. That is, until an enzyme in a feline retrovirus was discovered in 1970 that catalysed the manufacture of DNA from RNA. It was therefore called **reverse transcriptase**. The more accurate term is 'RNA-dependent DNA polymerase'. Retroviruses convert their single stranded RNA genome into a single stranded DNA copy so that it can be incorporated into the host cell nuclear DNA (termed **proviral DNA**) and thence undergo transcription as per normal. Once again, the viral genes code for structural capsid proteins, and these are packaged along with transcribed viral RNA and new reverse transcriptase.

Retroviruses therefore involve both nuclear and cytoplasmic machinery in their replication cycle. As with negative strand RNA viruses, the RNA alone is useless as a replicative agent.

**Main types of virus genome**
double stranded DNA viruses: pox viruses, Herpes viruses
single (+) strand RNA viruses: poliovirus, rubella virus
single (−) strand RNA viruses: influenza virus, measles virus
RNA retrovirus: HIV

### ■ 3.5.6 VIRUS ASSEMBLY

Viral components are assembled in an ordered manner rather than relying on the random assembly of the subunits. For example, as proteins are built into the capsid, they expose a binding site for the next protein to attach. The sequential series of steps is called the **assembly pathway**. Table 3.1 summarises the sites of virus manufacture

**Table 3.1 Sites of manufacture of viruses according to their nucleic acid**

| Genome* | Self-replicating | Site of manufacture | Site of assembly |
| --- | --- | --- | --- |
| DNA | + | N | N |
| +RNA | + | C | C |
| −RNA | − | C | C |
| Retrovirus | − | C/N | C |

* The nucleic acid is considered by itself, i.e. without the addition of associated enzymes such as RNA polymerase or reverse transcriptase.

C: cytoplasm, N: nucleus

• **Figure 3.13** Replication strategy of a retroviruses. The viral DNA is incorporated into the host DNA from which it is transcribed into viral proteins and reverse transcriptase

**Retrovirus**

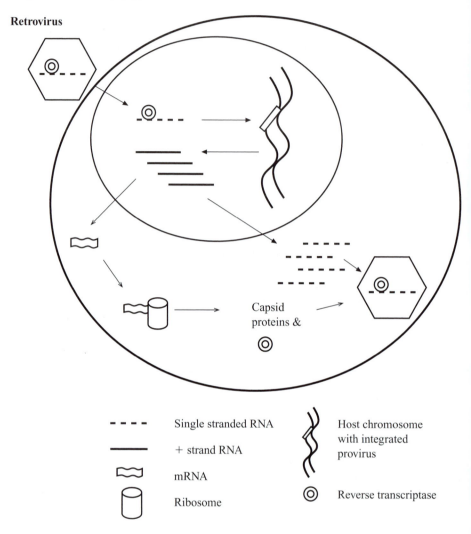

| | | | |
|---|---|---|---|
| - - - - | Single stranded RNA | | Host chromosome with integrated provirus |
| —— | + strand RNA | | |
| 〰 | mRNA | | |
| ⬚ | Ribosome | ◎ | Reverse transcriptase |

and assembly. The virus needs to exert some order on the sequence of manufacture of nucleic acid and structural capsid proteins, otherwise the host cell may degrade them as they accumulate. Indeed, it is puzzling how viral nucleic acid is protected from complete degradation by host nucleases. The sequence and quantity of the manufacture of components are two mechanisms by which viruses regulate their assembly. Viruses will tend to produce the enzymes they need in the appropriate order. Hence enzymes that are required early will be at the 'start' of the genome, i.e. read first. The transcription and translation of **early proteins** are usually non-structural but enzymatic. For example, a DNA virus will need DNA polymerase early on to permit the amplification of a viral DNA template. The new nucleic acid can be used both to act as instructions (through further transcription of other enzymes) and can be packaged up as a new virus. The genes coding for production of capsid (structural) proteins can be placed further downstream the viral genome. The transcription of these enzymes (late proteins) will then mean that capsid assembly follows nucleic acid replication. Clearly, manufacture of capsid before the viral genome is a more risky route. The loss of the genome, the sole source of all the viral information, is terminal whereas more capsid can always be made providing the genes are there. The quantity of each structural protein produced reflects that needed within the capsid and this will help order the sequence and speed of assembly.

### ■ 3.5.7 VIRUS RELEASE
Newly formed viruses are released from within the host cell by several mechanisms, depending on the virus type. Whilst effective, killing the cell will cause it to lyse and release all its contents. This, however, will terminate the production of the virus. Consequently, many viruses will attempt to exit in a non-traumatic manner. Enveloped viruses assembled in the cytoplasm can bud through taking the new envelope from the plasma membrane. **Budding** can be seen as the reverse process of membrane fusion internalisation shown in Figure 3.9. Enveloped viruses assembled in the nucleus can coat themselves with host nuclear membrane and then proceed to the plasma membrane.

A virus can spread and infect more cells if it causes the host cell to fuse with its neighbours. This property can be seen in certain virus-infected cells when cultured *in vitro*. One notable advantage is that the virus is protected from neutralising antibodies that may be circulating in the extracellular fluids.

The route of exit of the virus will also contribute to the pattern of spread of the virus. In virus-infected epithelial cells lining the respiratory tract, for example, it will serve the virus best if it is released from the apical membrane, i.e. that facing the lumen. The basolateral membrane will release the virus into the bloodstream/lymph which is not directly infectious, but it may be important, however, if the virus has a systemic phase or secondary replication site in another organ (e.g. muscles). The patterns of infection within the host (i.e. at the level of the body organs rather than at the cellular level) will be discussed in Chapter 9.

In this overview of the strategies of virus replication, it is easy to see how the virus has hijacked the host cell machinery in order to manufacture viral genome and protein. The extent to which the virus controls the host cell becomes more remarkable the more that is discovered. Indeed, many of the great features of the workings of the mammalian host cell have been uncovered by virologists. Examples include the discovery of genetic mutations, restriction enzymes, the production of interferon and reverse transcriptase.

### ■ 3.6 PROPAGATION OF VIRUSES
The term 'propagation' is preferred to 'culture' because it helps distinguish the mechanism of virus manufacture and assembly from the ordered increase in mass prior to binary fission of bacteria. The absolute dependence of viruses on the host cell for their replication dictates the absolute requirement of living cells or tissues for growing viruses in the laboratory. The culture of virus can be carried out using the living cells of increasing complexity up to intact organisms, i.e., cell culture, tissue culture, organ culture or eggs.

### ■ 3.6.1 CELL CULTURE
Cell culture is the use of a clone of a single cell type cultured in sterile flasks. Hundreds of cell lines are now available. Some grow as a monolayer on the base of a sterile plastic/glass flask whereas others grow as suspensions. The extent to which their

normal replicative lifespan is altered for each cell line provides a means for classifying cell lines into three types.

(a) Primary cell lines: these cells will divide only one or two times (more commonly referred to as one or two passages). The cells are diploid (two sets of chromosomes), as are most cells *in vivo*, and will survive, even if they do not divide, for a month or two. Primary cell lines are usually taken from animals and consequently suffer from cost and risk of contaminating viruses already present in the animal. Nevertheless, primary cell lines will often permit growth of viruses that continuous cells will not.

(b) Semi-continuous cells: these cells will undergo division for up to 50 generations before they senesce. They also are diploid. They are usually taken from foetal tissues like kidney and lung.

(c) Continuous cell line: these cells are either taken from a tumour or they have transformed spontaneously. They will replicate for well over 50 generations. In keeping with their abnormal (but extremely convenient) properties, they are aneuploid (abnormal number of chromosomes).

A list of cell lines widely used to culture viruses is given in Table 3.2. Knowing the chromosome patterns of cell lines has implications for viral vaccine production. When viruses are cultivated for use in vaccines there will be a requirement for them to be cultured in cells with normal chromosome numbers and normal division characteristics. The risk of transferring tumourigenic characteristics (as genes or cells) through using transformed cells (cells with tumour-like properties) is considered unacceptable.

It has been found that viruses that are found in the respiratory tract are recovered from specimens more readily if the cell lines are cultured at 33°C rather than 37°C. It is tempting to correlate this with the reduced temperatures found in the upper respiratory tract of people breathing cold air.

The medium used to grow and maintain the monolayers of cells will include a range of essential salts, and nutrients in a buffered solution. The medium will have growth factors supplied through the use of foetal calf serum (5–20 per cent v/v). This will promote the cells to divide within the flask.

### ■ 3.6.2 TISSUE CULTURE

Tissue culture is the maintenance of fragments of tissue in a flask. As the clump of cells multiplies then different cell types (usually fibroblasts) radiate away from the tissue. The nature and number of cell types are not necessarily known. This lack of exactness may not be a serious disadvantage providing the growth of virus is achieved. This practice is not used regularly because of the variability in results. Furthermore, the term is often used synonymously with 'cell culture'.

**Table 3.2 Typical cell lines used to culture certain viruses**

| Cell character | Cell type | Origin | Viruses for which they are used |
|---|---|---|---|
| Primary | kidney | Rhesus monkey | influenza, enteroviruses |
| Semi-continuous | embryonic lung | human | Herpes viruses |
| Continuous | HeLa | human cervical cancer | enteroviruses, adenoviruses |
| | Hep2 | human laryngeal | adenoviruses |
| | Vero | monkey kidney | measles virus |
| | Caco-2 | human colonic tumour | rotavirus |

## ■ 3.6.3 ORGAN CULTURE

Where possible the maintenance of a piece of an organ *in vitro* will provide a suitable representation of the *in vivo* process. The use of sections of trachea as tracheal rings will provide a good system with which to observe the inhibition of the cilia beating by influenza viruses. The problems of maintaining adequate oxygen supply to the centre of the organ section is usually a major difficulty in maintaining viability.

### 3.6.3.1 Eggs

Many viruses are able to grow when injected into embryonated hen's eggs. Fertilised eggs are incubated for up to 14 days at 37°C and then inoculated with the virus. The use of eggs has fallen from favour but still provides a neatly contained culture system, supplying fresh nutrients, free from any inhibitory antibodies. The different compartments within the embryonated chick egg will grow different viruses. For example, influenza virus grows in the amniotic fluid whereas pox viruses and herpes simplex will grow on the chorioallantoic membrane and produce lesions (**pocks**) that can be seen if the intact egg is illuminated from beneath.

### 3.6.4 Animals

Culturing viruses in animals is an expensive process but can be the most sensitive means of recovering virus from a patient. The use of newborn mice is the only reliable means of culturing some of the exotic viruses (Dengue, Ebola). Animals are also used for the production of antisera.

## ■ 3.7 DETECTING VIRUSES

Bacteriophages can be detected by inoculating a bacterial suspension with the phage and then culturing the bacteria on agar plate. Where the phage has multiplied in the bacterium there will be a zone of no bacterial growth, called a plaque. It is not possible to detect virus that infect humans using such an easy system because the viruses do not grow in bacteria but need particular eukaryotic cells. The plaque assay is of particular importance as it is used to assay the number of viruses in a sample. Serial dilutions of the sample are tested in the same manner as the serial dilution methods for estimating bacterial numbers. The number of plaques formed at each dilution are easily seen as gaps in the lawn of bacteria on a petri dish. The plaques are counted and the number of infectious virus particles is obtained by correcting for the dilution and sample volume.

---

### ■ BOX 3.3 PLAQUE COUNTS

For example, if 100 uL of the 1 in 1000 dilution yielded forty-two plaques then the original (undiluted) sample had $42 \times 1000 \times 10$ plaque-forming units (pfu) per mL. Formally this is:

$$\text{pfu/mL} = \frac{\text{number of plaques}}{\text{dilution} \times \text{mL}}$$

$$= \frac{42}{10^{-3} \times 0.1\,\text{mL}} = 4.2 \times 10^5\,\text{pfu/mL}$$

---

The number of infectious virus particles calculated from a plaque assay is not necessarily equal to the total number of virus particles present in the suspension because some virus particles are incomplete. The assembly of the virus will never be a

100 per cent efficient process and some particles may lack complete lengths of nucleic acid. If we counted the number of viruses by electron microscopy in the sample and obtained $2.0 \times 10^6$ virus particles per mL we can calculate the proportion of incomplete (non-infectious) virus particles. The ratio, the **efficiency of infection**, is:

$$\frac{\text{Total number of virus particles}}{\text{Number of infectious virus particles}}$$

$$\frac{2.0 \times 10^6/\text{mL}}{4.2 \times 10^5/\text{mL}} = 4.76$$

It is usual that the efficiency of infection $> 1$.

A further point concerning the propagation of phages *in vitro* is the ratio of phage to bacterium. If the phage is to undergo more than one round of propagation then there needs to be sufficient numbers of bacteria so that the phage does not destroy all the bacteria during the first replication. The ratio of phage to bacterium, which is called the **multiplicity of infection**, needs to be less than 1 otherwise there will be no discernable plaques on the bacterial culture because every bacterium has been lysed by the phage. A suitably low multiplicity of infection is roughly 0.01–0.1 infectious virus particles per bacterium; in this way individual plaques are obtained and several rounds of virus propagation can occur if large yields of virus are needed.

### ■ 3.7.1 HOW DO YOU KNOW IF THE VIRUS IS PRESENT WITHIN THE CELL CULTURE SYSTEM?

If a virus grows intracellularly and does not kill the cell one might expect that infected cells remain indistinguishable from uninfected cells. Fortunately, certain viruses betray their presence by causing such disruption to the cells that they cause visible **cytopathic effect** (CPE). This is a general term for any alteration in the morphology of the cells caused by the virus. Not all viruses induce cytopathic effects. The cytopathic effect caused by a particular virus is dependent on the cell line, so that viruses may be detectable in one cell type but not another. If a typical CPE is observed it will need to be confirmed by more specific tests. Examples of various cytopathic effects are shown in Figure 3.14, and discussed below.

- **Formation of inclusion bodies.** Unusual organelles may appear within virus-infected cells and are described as 'inclusion bodies'. They are thought to be accumulations of virus subunits and may develop in the nucleus (nuclear inclusion bodies) or in the cytoplasm depending on the site of replication of the virus. Often they are typical for a specific virus. Negri bodies, for example, are inclusion bodies that appear in the cytoplasm of neurons infected with rabies virus. Finding Negri bodies in a cell is a diagnostic laboratory finding for rabies virus.
- **Changes in the shape of the infected cells.** Herpes viruses (especially *Herpes simplex*) typically cause cells that are normally spindle shaped to round up and give a 'cluster of grapes' appearance.
- **Cell lysis.** The lysis of cells by a virus is a feature of several viruses when grown in cell lines. The pattern of the gaps left by the detached dead cells from the cell monolayer is called a 'plaque'. The number of plaques formed from a known volume of virus suspension can be used to count the number of virus particles. Assuming one virus causes one plaque, each virus is equivalent to one plaque-

• **Figure 3.14** Cytopathic effects found in various viral infections of cells. (1) Formation of inclusion bodies with either cytoplasmic or nuclear location. (2) Cell lysis. (3) Alteration in cell shape. (4) Fusion of cells. The last two examples, (5) Haemadsorption and (6) Immunofluorescence, are not directly visible cytopathic effects

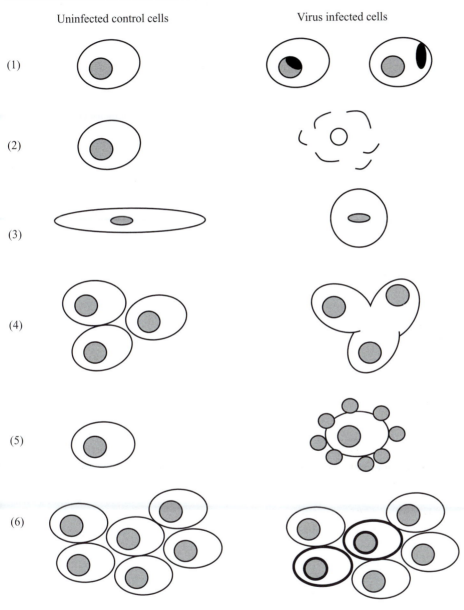

forming unit (PFU), a term that mirrors the colony-forming unit with bacteria cultures. Lytic viruses cause the cell to disrupt, typically when non-enveloped virus particles are released. Clearly this is fatal to the cell in question.

• **Cell fusion.** Infected cells may fuse and form multinucleate cells called **syncytia** when infected with certain viruses. Usually the cells will lyse under these abnormal circumstances.

• **Haemadsorption.** Some viruses give no morphological clues and need other procedures for their detection. Influenza virus has proteins on the end of spikes called haemagglutinins. They bind red cells. So, if the layer of cells is covered with red cells, the cells infected with influenza virus will bind the red cells and this is observable down the microscope following washing of the cells with saline to remove those unbound. This is termed **haemadsorption** rather than haemagglutination (clumping of a suspension of red cells).

- **Immunofluorescence.** An alternative means of detecting the virus in the absence of a cytopathic effect is immunofluorescence. A virus-specific antibody (raised in an animal) will bind virus antigen and if tagged with a fluorescent dye (typically fluorescein) will fluoresce under UV light. This technique will be limited to only detecting the virus you choose to look for. This is a plus in terms of specificity (you have identified the virus!) but if any other virus was present it will not pick it up. Consequently, you will not rely on this method to screen for a range of different viruses.

## ■ 3.8 OTHER MEANS OF DETECTING VIRUSES

### 3.8.1 ELECTRON MICROSCOPY

It is not always necessary to culture viruses in cell culture as a means of detecting them. One can visualise viruses directly by using an electron microscope. The light microscope cannot readily detect viruses because the resolution is not sufficient. Resolution is the ability to distinguish between two distinct spots that are close together. Light microscopes cannot resolve the gap between two points that are closer together than the wavelength of the light used to illuminate them. They appear as a single blurred spot. With the use of an electron beam, the electron microscope can resolve objects that are closer together than those distinguished by the light microscope. The electron beam has a shorter wavelength than the wavelengths of light (including ultraviolet light), hence electron microscopes can readily resolve objects as small as 2 nm, in contrast to light microscopy which can only resolve objects down to around 200 nm in size. Note that it is resolution and not magnification that is the critical feature in microscopy. Magnifying a blurred image only gives you a larger blurred image. Electron microscopes can supply two types of images: transmission electron microscopes (TEM) send the electrons through the very thin slices of the specimen, whereas scanning electron microscopes (SEM) only yield surface structures and at less resolution than TEM, but this is more appropriate for visualising intact virus particles in suspension.

Large viruses such as the pox viruses (e.g. the smallpox virus: *Variola virus*) are around 200 nm in size and visible by light microscopy, just.

Electron microscopy requires a level of expertise if small numbers of individual virus particles are to be detected in samples. A modification that improves the ease with which viruses can be spotted in a sea of debris is to first clump the viruses together by adding specific antibody to the sample. Collections of viruses held together by an antibody make them easier to spot. The technique is called 'immune electron microscopy'.

### ■ 3.8.2 ANTIBODY DETECTION

The laboratory diagnosis of virus infections is often made by identifying the presence of specific antibodies in the blood of patients produced in response to the infection. The sequence in which the class of antibody appears in the bloodstream is of diagnostic value. The immunoglobulin class that is produced first is IgM, followed by IgG; thus detection of IgM antibodies strongly indicates that the patient has a recent (or concurrent) infection. The methods for detecting antibodies usually employ anti-antibodies, raised in rabbits to detect the IgM. In other words, the IgM antibody in question is injected into a rabbit (i.e. human IgM acts as the antigen in the rabbit) and anti-IgM antibodies produced by the rabbit are extracted and used to detect the IgM in the patient's blood. To visualise the binding of the rabbit anti-IgM with the patient's IgM, an enzyme reaction is attached to the anti-IgM. The enzyme reacts with a suitable substrate that forms a coloured product for measurement spectrophotometrically. This type of assay is best represented by the ELISA (enzyme-linked immunosorbant assay) method. One problem with using antibody tests as diagnostic markers is the need to wait for the patient to mount an immune response in the first place. If the patient has a life threatening viral infection, there is urgent need to diagnose the condition immediately. Waiting 5–7 days for an immune response could be fatal. This is one reason why diagnostic virology laboratories are keen to use molecular methods to detect viral nucleic acid directly in the patient sample.

### ■ 3.8.3 ANTIGEN DETECTION

Antibodies that recognise viral antigens can be used to directly bind to viral antigens expressed in infected cells. This can be used to diagnose specific virus infections in samples from the patient. A fluorescent label such as fluorescein can be attached to the antibody as a means to visualise the virus.

### ■ 3.8.4 NUCLEIC ACID DETECTION

In addition to delays inherent in serological diagnoses, many viruses have not been successfully cultured in cell culture (e.g. hepatitis B virus). To obtain quicker results for diagnosis the older methods have been supplanted by direct detection of viral genome with nucleic acid probe methods. The methods rely on the hybridisation of a specific sequence of nucleic acid (the probe) to the target nucleic acid in the sample followed by a means of visualising the hybridisation (e.g. a fluorescent label). Molecular probe techniques will either bind directly to the target sequence or, alternatively, the target is copied (amplified) first by a polymerase chain reaction. However, amplification runs the risk of amplification of contaminating material and the estimation of the original numbers of microbe can be difficult.

### ■ SUMMARY

Viruses infect all living organisms, including bacteria. The nucleic acid is housed within a helical or (more commonly in human viruses) icosahedral protein shell called a 'capsid'. The nucleic acid is either RNA or DNA, but never both. The capsid shell and nucleic acid (collectively termed the 'nucleocapsid') may be surrounded by a phospholipid envelope but such viruses are susceptible to drying in the environment. Viruses can be classified by their physical properties (nucleic acid, presence of envelope) and the replication strategy. Viruses only replicate within host cells and, hence, are inert in the environment. Viral replication follows a series of steps: attachment, internalisation, genome replication, assembly and release. The type of nucleic acid (single stranded, double stranded DNA or RNA) determines the replication strategy for the virus. Viruses are assembled from subunits rather than grown (increased in mass), hence viruses have a distinct 'growth' curve from bacteria. As strict intracellular parasites, viruses are 'cultured' in cell culture, tissue culture systems or in eggs and animals. The replication of virus may cause morphological alterations to the infected cells (cytopathic effects) that are often characteristic of the particular virus.

## RECOMMENDED READING

Cann, A.J. (2001) *Principles of Molecular Virology*, 3rd edition, Academic Press, London, UK.

Collier, J. and Oxford, J. (2000) *Human Virology*, 2nd edition, Oxford University Press, Oxford, UK.

Dimmock, N.J. and Primrose, S.B. (1994) *Introduction to Modern Virology*, 4th edition, Blackwell Press, Oxford, UK.

White, D.O. and Fenner, F. (1994) *Medical Virology*, 4th edition, Academic Press, London, UK.

## REVIEW QUESTIONS

*Question 3.1*  How do the growth curves obtained *in vitro* for viruses and bacteria compare?

*Question 3.2*  Is viral nucleic acid alone infectious?

*Question 3.3*  What is the significance of the icosahedron for viruses?

*Question 3.4*  How does the efficiency of infection correspond to total and viable counts for bacteria?

# FUNGI

# 4

Fungi comprise the untidy collection of micro-organisms that includes moulds and yeasts. Toadstools and mushrooms are actually just particular structures formed by moulds and not separate organisms. Whilst fungi are of great benefit to mankind and play an essential recycling role in nature, they can produce disease in humans, hence our interest. Of the micro-organisms considered in this book, fungi are the only organisms that have forms that are visible to the naked eye. In addition to mushrooms and toadstools, the filamentous collections of threads that characterise moulds in the bread bin are also readily seen by eye. Single yeast cells, however, are microscopic despite being up to 10 times the size of bacterial cells.

The fungal lifestyle can be viewed as two phases:

- the **trophic** (feeding) phase, mediated through radiating mycelia, and
- the **dispersal** phase, mediated through production and release of spores.

Whilst termed micro-organisms, fungi attempt to monopolise their potentially large size. If nutrients exist, they will continue to increase in size, with no restrictions on the final dimensions. In terms of size, it is estimated that certain fungi, as mycelia radiating in soil, are amongst the largest living organisms existing and can be considered to represent a primitive multicellular organism. Like true multicellular organisms, fungi produce specialised structures with distinct functions, notably the production of 'sexual organs' and large fruiting bodies. Mushrooms are a class of fungus that produce such fruiting body structures that can be eaten. Other than producing poisons which are toxic to man if eaten, mushrooms do not directly cause infections of man and will not be central to our discussions. Likewise, the Brewer's yeast, despite the not insignificant property of converting sugars into alcohol and carbon dioxide, will not be discussed further.

## ■ 4.1 FUNGI: MOULDS AND YEASTS

Fungi are eukaryotes, typified by the possession of a nuclear membrane and mitochondria. Unlike bacteria, fungi are usually multicellular organisms which possess complex lifecycles. Whilst ubiquitous in soil and many varied environments, fungi are less tolerant of adverse conditions than bacteria and prefer warm, moist conditions.

## ■ 4.1.1 MOULDS

Moulds are filamentous fungi; that is, they exist as rigid-walled filaments. These filaments are a mass of branched tubes called **hyphae** which collectively are called a **mycelium**. The mycelia permit great variation in the structures of the fungus, as the production of a small bud that protrudes from the main cell body and increases in size until it is pinched off, leaving a scar on the parent cell. Fission yeasts divide through a binary fission process in which the yeast divides in two with the production of a septum. The growth of the yeast may not simply be through multiplication of more and more yeasts but they may produce hyphal outgrowths:

1. **germ tubes**, hyphal protuberances that emerge from the yeast cell. The germ tube can be considered as an immature form of hyphum, i.e. before septa have appeared.
2. Under certain conditions yeast can form **pseudohyphae**, chains of young (small) yeast cells formed by budding but still attached. Pseudohyphae appear constricted at the junctions between the attached cells whereas true hyphae show no constriction at the septal join between cells.

Some yeasts are diploid and reproduce sexually via fusion of different **mating types**. Those yeasts that do not reproduce sexually are termed **anamorphic yeasts** (see Box 4.1).

Yeasts are different from moulds in several respects.

- The genome is smaller, reflecting a narrower metabolic capacity.
- Yeasts do not form secondary metabolites.
- Yeasts do not form the variety of morphological structures of true fungi.
- Yeasts are less tolerant of adverse conditions (pH, temperature, etc.).

## ■ 4.2 FUNGAL STRUCTURE AND FUNCTION

### 4.2.1 CELL WALL

Fungi have rigid cell walls (like that of bacteria) that determine the shape of the fungal structures but the chemical composition and arrangement is different to bacteria. Fungi do not possess peptidoglycan and therefore are not susceptible to antibiotics such as penicillins. The fungal cell wall is comprised of a greater variety of substances and the number of layers is also greater. Most characteristic is the use of the polysaccharides **chitin** and **glucan** in the cell wall. Neither glucan nor chitin are present in mammalian or bacterial cell walls. Chitin is an extremely tough, linear (homo)polymer of **N-acetyl-glucosamine** and is predominant as bundles which strengthen the cell wall. Glucan is a linear homopolymer of glucose units which are connected by 1–3 and/or 1–6 linkages (the numbers referring to the carbon number in the carbohydrate molecule). Both are insoluble in water and provide rigidity to the cell wall. Whereas bacteria cross-link the peptidoglycan with peptide strands, fungi cross-link with the glucan chains. The other typical fungal wall constituent is a variety of **mannoproteins**. The composition of

---

**■ BOX 4.1 SEXUAL AND ASEXUAL FORMS OF REPRODUCTION**

Two terms that should be understood are **anamorph** and **teleomorph**. Because fungi may have a sexual and/or asexual means of propagating itself there have been two taxonomic names possible for the same organism. The anamorphic state refers to the organism named according to its asexual state, whereas teleomorph is the name of the sexual form. A third term, **holomorph**, can be used for the name of a fungus in which both states have been recognised and named. In that case, the name of the sexual state takes priority.

• **Figure 4.3** A diagrammatic representation of the fungal cell wall. The cytoplasmic membrane is included so as to demonstrate the presence of a cytoskeleton. The cell wall itself is essentially of two layers: chitin and cross-linked glucans, and a mannoprotein layer; that is, covalently bound to the glucans

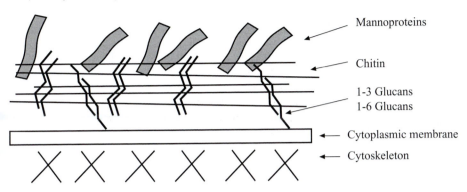

Mannoproteins

Chitin

1-3 Glucans
1-6 Glucans

Cytoplasmic membrane

Cytoskeleton

fungal cell walls is dominated by 80–90 per cent polysaccharide. Figure 4.3 shows a diagrammatic representation of the key fungal cell wall components. Note that beneath the cytoplasmic membrane (in the cytosol) the **cytoskeleton** is indicated. The cytoskeleton, which includes actin, acts as a scaffold inside the cytoplasmic membrane and helps in the assembly of new cell walls by directing the vesicles to the apical tip.

The apical growth that characterises growth of mycelial fungi differs from that seen in yeasts and bacteria (Figure 4.4). Whereas fungi demonstrate apical growth, bacteria grow from the centre of the bacterial cell. Yeasts grow either by budding or (like bacteria) by fission. Apical growth means that the tip needs to be sufficiently plastic to permit the insertion of new cell wall material. The new cell wall is incorporated into the tip via fusion of the dense collection of vesicles at the advancing tip, that collectively are known as the **Spitzenkörper** (or **apical body**). The fungal cell wall becomes more rigid along the trunk of the hyphum as the 'plastic' tip advances progressively. The trunk becomes rigid because of increasing cross-linking of the glucan strands and crystallisation, a particular arrangement of the strands of chitin (Figure 4.5).

### ■ 4.2.2 SPORES

The collections of spores produced following sexual fusion of male and female hyphae in different fungi have characteristic morphological patterns which are useful for classification purposes. Although all spores themselves are produced by an asexual process, it is not uncommon to find references to sexual and asexual spores, depending on the sexual pattern observed prior to the production of spores.

Although the spores produced by fungi are useful to mycologists in the classification and identification of fungi, their role for the fungus lies in their distinct function:

- asexual spores are produced following mitosis and are produced to increase the mass of fungus produced (when the spores germinate into hyphae),
- large numbers of small sexual spores are produced following meiosis and produced on specialised structures that promote the dispersal of the spores on to new environments.

In order to disseminate spores into the environment, fungi employ the wind to carry them. This distribution mechanism will work best if the spores are elevated in some manner, so as to catch the wind. The hyphae that can be found extending aerially are

• **Figure 4.4** Patterns of cell growth. Bacteria, yeasts and fungi differ in the sites at which new cell wall is incorporated. Bacteria (A) deposit new cell wall along the length of the cell wall. When the cell starts to divide, the new cell wall is deposited at the site of division (the septum). In contrast, fungi (B) grow by incorporation of new cell wall only at the tips. Yeasts grow by either fission (C) or by budding (D). Note that fission is the same process in bacteria and yeasts

(A) Bacterium

(B) Fungal hyphum

(C) Yeast (fission)

(D) Yeast (budding)

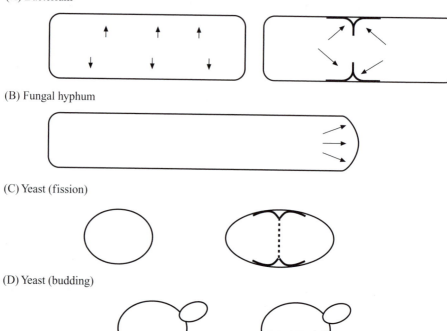

• **Figure 4.5** The apical growth zone of the fungal hyphum is shown with the vesicles (circles) supplying the new cell wall material. The structural arrangements for the cell wall at the tip and along the trunk are given. The tip is represented as a less ordered structure where the straight lines represent linear chitin chains and the thicker curved lines are the glucan chains. The grouped glucan chains in the trunk represent the increased cross-linking and concomitant rigidity

Trunk                                                        Tip

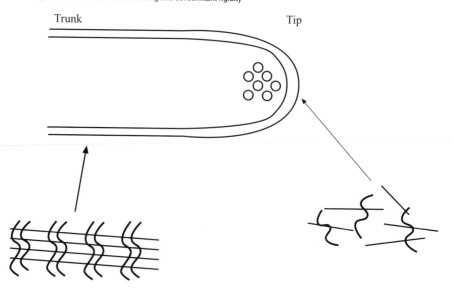

used for this purpose. The hyphae will often contain numerous spores at the tip and such a structure is called a **conidiophore** (see below, p. 116). **Conidia** are pigmented and impart the distinct colours that colonies of moulds show when cultured on agar plates. This strategy of releasing spores into the air may be successful for soil-living fungi and has significance in the airborne acquisition of certain fungi by humans breathing dusty air containing large numbers of spores.

The environmental conditions that the fungus finds itself in will influence the extent to which a fungus undergoes sexual or asexual reproduction. The type of nutrient (protein or carbohydrate) and its concentration are key modulators. In general, if nutrients are scarce, then sexual reproduction is favoured. The triggers may indeed be the unbalanced nature of the nutrition. Under conditions where nutrients are plentiful and growth of the hyphae is promoted, then asexual spore formation is usual.

## ■ 4.3 FUNGAL GROWTH AND METABOLISM

### 4.3.1 GROWTH

The key characteristic of fungi is the ability to produce mycelia through apical growth.

As the hyphae extend radially (outwards) across a surface, the fungus is able to colonise a large area. The rate of branching of the hyphae will depend on the concentration of nutrients. If the environment is nutrient poor, then the hyphae will increase in length rather than produce branches. In nutrient-rich environments, however, the hypae will branch more frequently in order to maximally utilise the available nutrients (Figure 4.6).

The growth of fungi in laboratory cultures is largely an artefact, in the same way that culturing bacteria in laboratory culture media has little resemblance to the natural environment. With this proviso, let us consider some of the features of fungal growth *in*

• **Figure 4.6** Hyphal extension in nutrient-rich and -poor environments. A fungus will regulate the proportions of hyphal length to number of branches so as to best find and exploit the available nutrients. This ratio of hyphal length and tip number is known as the 'hyphal growth unit'

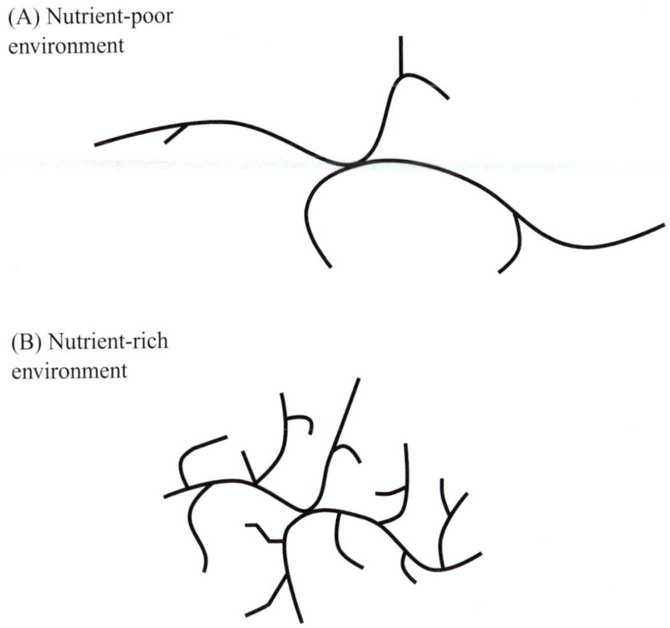

(A) Nutrient-poor environment

(B) Nutrient-rich environment

*vitro*. The growth of yeasts and hyphal moulds in liquid batch culture resembles the increase in numbers and mass seen in bacterial growth. The same phases of growth are seen: lag, exponential, stationary and decline. The resemblance is best during the early stages of the growth, when the entire fungus is growing. As hyphal growth occurs at the tips only (**apical growth**), the centre of the mycelial mat will autolyse (no viable fungus can be recovered from the centre). The increase in fungal mass occurs only when all hyphae continue to branch outwards and this occurs in the early stages of the culture only. When fungal moulds are cultured in broth cultures, they tend to form spherical balls which will restrict the flow of gases and nutrients into the centre of the fungal ball. Under these conditions the hyphae will continue to grow and branch at the surface and not in the centre. The mass of the fungus will be proportional to the volume of the sphere ($4/3\ \pi r^3$). From this it can be shown that the cube root of the mass is proportional to time. This is called 'cubic growth' and means that instead of a graph of log mass versus time, the cube root of the mass of the fungus over time is needed to get the straight line graph.

Fungi will also tend to grow as a mat on the surface of liquid broth media if they are not vigorously shaken. The increasing thickness of the fungal mat will progressively reduce gaseous exchange in the broth. Shaking will lead to the formation of fungal balls. In effect, 'problems all round', if you will pardon the pun.

The mycelium will increase in size until the fungus has depleted the nutrients from its surroundings. The suboptimal growth rate signals a switch from primary metabolism to **secondary metabolism**. The products are termed 'secondary' because they are not involved in the growth of the organism rather than being produced after the growth phase. As with bacteria, the products of secondary metabolism are not utilised for growth of the organism but instead serve to aid its survival once nutrient concentrations have become critical. One of the important functions of secondary metabolites is to trigger the formation of sporulation in order to shed spores into new territory. Antibiotics are also secondary metabolites and they function as a survival mechanism by acting on susceptible organisms in the vicinity that may compete with the organism for the nutrients. You may notice that of the two functions described, one acts on the host organism whereas the other acts on competitors.

Whilst many of these features mean that many fungi are able to attack a diverse range of substrates (wood, leather, cloth) the relevance of these attributes to infecting man is not clear. It might be expected that fungi would cause a significant number of infections in man rather than existing predominantly in the environment as soil- and plant-associated organisms. The conditions necessary for fungal infection in man are considered in Chapter 10.

## ■ 4.3.2 METABOLISM

Fungi are distinct from plants because they can grow without light. Hence, fungi do not possess chlorophyll and cannot use carbon dioxide as a carbon source. They therefore need to obtain their energy through direct uptake of non-living organic matter (heterotrophic or, to be more exact, chemoorganotrophic). The uptake will need to be of low molecular weight and in a soluble form because, unlike most other eukaryotes, fungi bacteria possess a rigid cell wall and therefore cannot import food via endocytosis. Dependence on pre-formed sources of nutrients means that fungi have adopted a lifestyle that absorbs food from either

- non-living (dead) organic matter, or
- living organic matter.

Fungi that obtain nutrients from their host (and at the host's expense) are called 'parasitic fungi'. Some fungal associations with their host are of benefit to both fungus and host, i.e. **symbiotic**. For example, many plants, especially orchids, cannot grow unless the fungi have colonised their root system (**mycorrhizal fungi**).

Fungi have limited nutritional requirements. They need carbon and nitrogen in an organic form. Carbon is abundant in nature but may be present in a form that is unavailable to most other organisms, notably as polymers such as cellulose, chitin and keratin. Carbon in cellulose is the most abundant atom on the planet. We on Earth would be overrun with discarded plant material were it not for the ability of fungi to break down plant cellulose in the cell wall of plants. Different fungi have different nutritional requirements (reflecting their requirements for nitrogen and carbon in specific chemical forms) and this will determine where the fungi are to be found. With humans, fungi are unique in their ability to extract carbon and nitrogen from the keratin in dead skin. The fungi secrete **degradative enzymes** into the environment that will hydrolyse the polymers: cellulases, chitinases and keratinases. Fungi will readily grow with glucose as the carbon and energy source but it is unlikely that in their natural environments it is readily available. The nutrients freely diffuse across the rigid cell wall but will be actively transported into the cytosol across the cytoplasmic membrane.

The waste business appears to be profitable. As decomposers and recyclers, fungi appear to be doing very well. There are tenfold more fungal species currently recognised (c.90,000) than bacteria (c.4000) and viruses (c.6000). All three microbial types are thought to be woefully underestimated and the fraction of possible fungal species currently recognised is possibly as little as 10 per cent of the real number.

Most fungi are essentially aerobic but are able to tolerate anaerobic conditions. With sufficient oxygen available, fungi can continue to grow. Without oxygen, growth ceases and aerobic metabolism is switched to fermentation to maintain sufficient energy for basal metabolic activities. Yeasts are best suited to dealing with anaerobic fermentation as all beer drinkers are aware. The production of ethanol and carbon dioxide from carbohydrates by brewing yeasts is likely to be the most well known biochemical pathway. The anaerobic fungi are grouped in with the lower fungi and do not have any direct dealings with humans.

Typical conditions for fungal growth are humid atmosphere, damp surface (soil) with a moderate temperatures (between 4°C and 20°C). Yeasts are able to tolerate situations of low free water ($a_w$) and they can typically be isolated from fruit. Ripe fruit has high concentrations of solutes such as alcohol and sugars which reduces the available water.

## ■ 4.4 FUNGAL GENETICS

The fungal genome (i.e. the total genetic information) consists of the chromosome and mitochondrial DNA along with extrachromosomal elements: plasmids. The former two elements are consistent with other eukaryotes whereas the possession of plasmids is a feature of prokaryotes.

Fungal genomes are not that much larger than those of bacteria and, consequently, are small compared with other eukaryotes. Fungi do not contain multiple copies of the same genes, nor do they have much nonsense (i.e. non-coding) DNA (called **introns**).

Most fungi are **haploid** (possess one set of unpaired chromosomes, N), although the medically important yeast *Candida albicans* is **diploid**; that is, the chromosomes are in pairs (2N), each pair being homologous.

Sexual reproduction in moulds is a vast topic because the different classes of fungi reproduce by different mechanisms and the sexual organs have separate names. In general, lower fungi possess specialised sexual organs whereas the higher fungi have relatively inconspicuous organs, if at all. Many of the fungi that infect man are members of the **fungi imperfecti** and therefore the sexual stage has not been identified. For those reasons you might find many mycology texts somewhat unforgiving.

Both sexual and asexual reproduction can be found in most fungi, although many will predominantly use asexual reproduction and only rarely use sexual means. Some fungi, often those that cause infections in man, appear to have completely lost the ability to reproduce sexually. The massive numbers of species of fungi indicates that fungi have benefited from using both sexual and asexual methods of reproduction.

As a mechanism to prevent genetically identical cells from fusing, many fungi have developed mating systems whereby cells of the same **mating type** cannot fuse. Such a system resembles the role of male and female by forcing genetic diversity through recombination of compatible nuclei. Many higher fungi contain nuclei of two mating types (a male and female equivalent) and are thus able to undergo sexual reproduction from the same fungal colony, in other words, they are self-fertile (can mate with themselves). Such fungi are called **homothallic**. Those that require fusion of nuclei of the opposite mating type are called **heterothallic**. Any such restriction on the ability of fungi to mate is called **incompatibility**. Incompatibility serves to promote genetic exchange (heterothallism) or to promote the growth of a particular phenotype by restricting outbreeding (homothallism). The role of mating is also considered next.

Sexual reproduction involves the cyclical alternation between **haploidy** and **diploidy**. The means of reducing the numbers of sets of chromosomes from diploidy (2N) to haploidy is meiosis. Meiosis consists of two divisions of the nucleus but the chromosome only dividing once, hence the sets of chromosomes are halved. For the fungus to produce a diploid cell it needs to fuse two haploid cells (**gametes**). The fusion of two sexually compatible cells (be they spores or hyphal cells) is called **plasmogamy**. The resulting cell now has two independent and non-identical nuclei, one from each parent/gamete. The subsequent fusion of the two nuclei is called **karyogamy** and leaves the cell with a diploid set of chromosomes ready for reduction by meiosis.

Once the binucleate state has been created by plasmogamy, the fusion of the two nuclei need not occur immediately, but the cell may multiply further with each of the daughter cells having two separate nuclei (N+N). Each pair of nuclei in a cell is termed a **dikaryon** (pair of non-identical nuclei).

Thus far we have described three critical and essential events:

- **plasmogamy**,
- **karyogamy**,
- **meiosis**.

Variation in the proportion of time fungi spend as haploid, diploid or dikaryotic cells helps explain the number of lifecycles that fungi have adopted (and help to explain why most mycology textbooks appear to describe endless impenetrable lifecycles with no obvious pattern!). We can now predict a number of lifecycles, which differ in the time spent as haploid, diploid or dikaryotic:

1. **Asexual lifecycle**. No sexual phase.
2. **Haploid lifecycle** (Figure 4.7). The vast majority of the cycle is haploid except for the short period necessary for fusion of two haploid nuclei to generate the diploid state which then promptly undergoes meiosis.
3. **Diploid lifecycle** (Figure 4.8). Predominantly the fungus is diploid with only the temporary phase of haploidy following meiosis.

• **Figure 4.7** The sequence of events for a haploid fungus

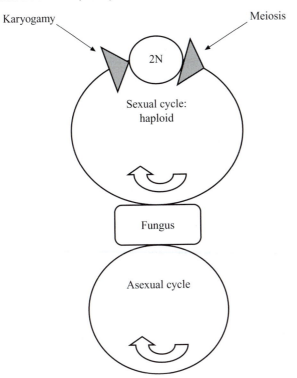

• **Figure 4.8** The sequence of events in the lifecycles of a diploid fungus

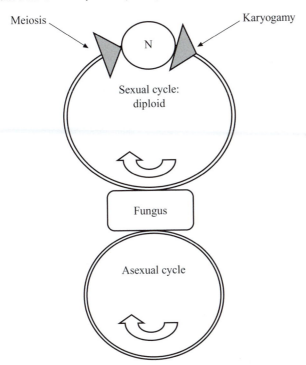

4. **Dikaryotic lifecycle** (Figure 4.9). The fungus forms dikaryotic cells by immediately fusing the haploid cells resulting from meiosis.

Further variations are recognised, e.g. haploid–diploid lifecycle, haploid–dikaryotic lifecycle, etc. Interestingly, lifecycles generally tend to be more complicated in the higher fungi, and the higher fungi tend to exist in more specialised ecological environments, although this does not appear to be the case with many fungi pathogenic to man.

### ■ 4.4.1 FUNGAL GENETICS: WHY SO COMPLICATED?

It is interesting to consider why fungi, being eukaryotes, are mostly haploid whereas eukaryotes are typically diploid, i.e. carry two sets of chromosomes. Diploidy is a mechanism whereby any mutations that occur in any one gene need not necessarily be expressed. If the mutant gene is recessive, it will not be expressed because the other, non-mutated copy is dominant. The recombination events that occur in sexual reproduction via meiosis may result in a mutation becoming expressed, but this will occur only if a similar recessive mutation is present in the other sexual mate. In contrast, mutations in haploid organisms will be expressed because there is no second copy to cover the altered gene. The advantage of the immediate exposure of mutations to haploid organisms is that any advantages the altered phenotype confers is immediate. The selective pressures on the organism either favours the mutant or eliminates it. In this way mutation in haploid organisms is of the moment, whereas diploid organisms can accumulate mutations for variation later. Consider then the fungi that possess coenocytic mycelia. They have more than one nucleus in the cytosol. If, then, a mutation occurs in one of the nuclei, the variation is hidden until such times as the fungal

• **Figure 4.9** The sequence of events in the lifecycles of a dikaryotic fungus

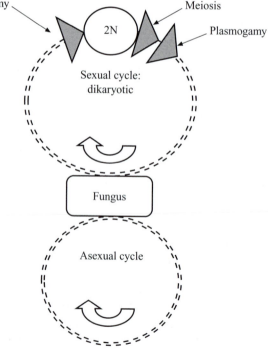

Karyogamy

Meiosis

2N

Plasmogamy

Sexual cycle: dikaryotic

Fungus

Asexual cycle

mycelia produce uninuclear spores. In this way moulds are haploid but retain some of the features of diploidy. As mentioned earlier, whilst the yeast *Candida albicans* is permanently diploid, most yeasts are haploid but, being separate cells, they cannot exploit the benefits of being coenocytic.

Fungal spores are produced following sexual mechanisms in response to adverse conditions, whereas asexual spores are used to enable the fungus to quickly multiply and exploit favourable conditions. Why is sex the reason for the distinct functions of sexually and asexually produced spores? Fungi are eukaryotes that have sexual and asexual mechanisms of reproduction. There are evolutionary advantages for both. Asexual reproduction is clonal in effect (more of the same is produced) and quick (mitosis is a faster process than meiosis and you don't have to wait for a partner); hence, if your existing genotype is working well in your current environment then you have little need to change it. Sexual reproduction in fungi involves the fusion of + and − mating type nuclei and meiosis. The genetic recombination can result in the creation of new phenotypes as a consequence of crossing-out errors: different combinations of alleles from the parental genomes. Meiosis, then, is a cause of genetic diversity and this mix and match effect on the genotype/phenotype will enable the fungus to produce variants that are suited to changing environmental conditions.

In addition, you may encounter the term **parasexuality** when reading about fungal genetics. Parasexuality is a cover-all term for further genetic tricks exhibited by fungi which are non-meiotic in mechanism but their objective remains the same: to generate genotypic diversity. One should note that most of them have only been observed under laboratory conditions.

## ■ 4.5 FUNGAL CLASSIFICATION

The taxonomic history of fungi is long and complicated. This is mentioned so as to warn readers that older texts are very difficult to cross-reference with each other because the classification varies considerably. Fungi are eukaryotic (they possess a nuclear membrane) and are placed as a distinct taxonomic family (Fungi) but historically have been grouped with plants. Fungi differ from plants in two key features:

- fungi are absorptive heterotrophs and lack chlorophyll and are thus non-photosynthetic, whereas plants are autotrophic, utilising photosynthesis via chlorophyll,
- fungal cell walls contain chitin but not cellulose.

In 1969, Whittaker rescued fungi from botanists by placing them as one of the five kingdoms (animals, plants, fungi, protists and monera). The diverse range of forms once considered to be fungi is reflected by the division of 'Fungi' into three kingdoms:

- Fungi,
- Chromista (water moulds and algae),
- Protozoa (slime moulds and chytrids).

Of these, the kingdom Fungi contains the organisms that are of interest to medical microbiologists and will be considered in this book. As discussed in Chapter 6, the descending ranking order in taxonomy is as follows: Phylum, Class, Order, Family, Genus and Species, so the three groups (Fungi, Chromista and Protozoa) are a long way from the levels Genus and Species, familiar to bacteriologists. A more selective and strict system of classifying fungi is emerging that should make a break from old groupings of algae, protozoa, etc. Indeed, the classification of fungi is likely to see a period of considerable dispute as newer genetic methods are employed. Traditionally, fungi have been classified according to the older, morphological character-based ideas. The features of the mycelia, the spores and sporangia and the sexual and asexual patterns of reproduction have resulted in a classification scheme consisting of six phyla. Fungi are described as lower and higher fungi in order to reflect evolutionary history, the lower fungi being earlier and more primitive forms. The lower fungi are those in the phyla:

- Zygomycota,
- Oomycota,
- Chytridomycota.

The higher fungi are represented in the phyla:

- Ascomycota,
- Basidiomycota, and
- Deuteromycota.

'Yeast' is not a formal taxonomic term but a description of a morphology, like the term 'mushroom'. Yeasts can be produced by fungi of different groups and genera, most commonly under anaerobic conditions. There are yeasts that do not appear to have any other morphological form because they have lost the ability to produce mycelia.

Most fungi produce spores that are generated following asexual reproduction (asexual spores) and following sexual reproduction (sexual spores) and important examples of the names given to the spores produced in different cycles by different fungi are given in Figure 4.10. For identification purposes these spores are critical. The structures that support or house these spores are also used to help identify fungi.

Asexual spores are found:

- in a sac called a **sporangium** which is supported by a **sporangiophore**. These spores are called **sporangiospores**. This pattern is typical of the lower fungi in the Class Zygomycota.

or

- unenclosed and simply projecting from a supporting structure called a **conidio-phore** and hence are called **conidia**. Conidia are typical of the other (higher) fungi in the Classes Ascomycota, Basidiomycota and Deuteromycota.

The sexual spores produced following meiosis are identified by the structures in which they are kept. Each Class of fungus can be identified by the type of sexual spore produced:

- Zygomycota produce zygospores.
- Ascomycota have sexual spores in special sacs called an ascus, hence the term 'ascospores'.

• **Figure 4.10** Fungi produce different structures to house spores depending on whether they undergo a sexual or asexual cycle of division. These are used to classify and identify the individual organisms

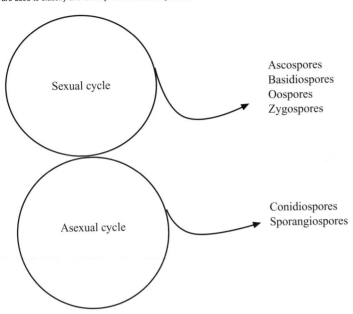

- Basidiomycetes produce basidiospores in which the spores are found on structures called a 'basidium'. Morphologically this resembles the overall structure seen in conidiospore production.
- The Deuteromycota are those fungi that do not produce sexual spores (hence Fungi Imperfecti).

Unfortunately, the more recent grouping of the fungi (*sensu latu*) has the Kingdom Fungi containing the phyla Ascomycota, Basidiomycota, Zygomycota and Chytridomycota. The inclusion of organisms with flagella (Chytridomycota) doesn't help unify the characteristics that constitute a fungus. In addition, the phylum Deuteromycota, the one collection of organisms of most clinical relevance, has been lost. The Deuteromycota, also called the **Fungi Imperfecti**, is a resting place (a euphemism for dumping ground) for those fungi in which the sexual stage has not been identified or perhaps does not exist. Of all the fungal groups it is an important group for medical microbiologists despite its uncertain provenance, in that several genera that cause human infections are grouped into this Class, in particular the genera Trichophyton, Microsporum and Epidermophyton. These are the genera that cause skin, nail and hair infections of man. With the newer molecular phylogenetic studies of micro-organisms, whereby comparisons can be made on evolutionary conserved gene sequences such as ribosomal subunit genes, it seems likely that the class Deuteromycota will become redundant. Even for species in which only the asexual form has been described, it should be possible to identify the sexual form using molecular comparisons.

## ■ SUMMARY

Fungi are characterised by apical growth through rigid tubular hyphae which accumulate to form a mycelium. Fungi in the trophic phase are concerned with feeding through the filamentous mycelia but for the colonisation of distinct, new environments fungi use spores for dispersal (dispersal phase). As saprophytic heterotrophs, fungi are important recyclers of nutrients from dead and decaying matter through excreting a diverse range of extracellular enzymes. This lifestyle matches the aerobic existence in the soil. Fungi resist unfavourable conditions through production of spores. As the centre of the fungal mycelium senesces, fungi need to periodically undergo recombination events, often sexual, but not always. Incompatibility systems prevent the co-existence of genetically distinct nuclei but other fungi possess incompatibility systems that do the opposite (i.e. force recombination through incompatibility between genetically identical mating types). All this complicated genetics means that fungi possess a variety of patterns in which the haploid and diploid phases of the cell cycle are organised. Classification and identification of fungi, particularly moulds, is based on the type of asexual and sexual spore pattern. Nevertheless, as many fungi are able to form both sexual and asexual spores, the accurate identification of the spore type depends on knowing whether the spores you see are produced following sexual or asexual processes. For example, fungi that cause ringworm are members of the genus *Trichophyton* and are in the Class Deuteromycota. However, a sexual form of this fungus has been identified and is classified as a member of the Ascomycota (*Arthroderma*) because of the ascospores that are found. The sexual form of a fungus is called the 'teleomorph', the asexual equivalent the 'anamorph'.

## RECOMMENDED READING

Carlisle, M.J., Watkinson, S.C. and Gooday, G.W. (2001) *The Fungi*, 2nd edition, Academic Press, London, UK.

Deacon, J.W. (1997) *Modern Mycology*, 3rd edition, Blackwell Scientific Publications, Oxford, UK.

Griffin, D.H. (1994) *Fungal Physiology*, 2nd edition, Wiley-Liss, New York, USA.

Guarro, J., Gene, J. and Stchigel, A.M. (1999) Developments in fungal taxonomy. *Clin. Microbiol. Rev.* 12; 454–500.

Jennings, D.J. and Lysek, G. (1999) *Fungal Biology*, 2nd edition, BIOS Scientific Publishers Ltd, Oxford, UK.

## REVIEW QUESTIONS

Question 4.1    How do moulds differ from bacteria in their growth in broth culture?

Question 4.2    Why are there four functional sections of a fungal hyphum?

Question 4.3    List the differences and similarities between fungi and plants.

Question 4.4    What are the Imperfect fungi?

# MICROBIAL DEATH

## ■ 5.1 INTRODUCTION

As far as is known, bacteria do not die of old age. They replicate themselves by division once they reach a certain size. If, in the process of duplication of the DNA lethal mutations are made, then the bacteria will cease to multiply and die. Alternatively bacteria die when conditions for survival exceed the normal conditions of growth, such as dehydration or exhaustion of nutrients. Bacteria vary in their ability to resist adverse conditions and the best example of a survival strategy is the bacterial endospore. Bacterial spores also vary in their resistance to dehydration, heat or freezing but in general they provide a very efficient mechanism for survival.

An understanding of how to kill microbes is clearly of great value. A few seconds' thought will reveal that knowledge on how to kill microbes is not just important in trying to treat human infections. The wider public health arena is important: water supplies and food products, for example. Also implicated are the veterinary sciences, food manufacture, brewing industries, preservation of materials such as building timber, and paper.

One difficulty that undermines our confidence in controlling bacteria and viruses is the definition of death in microbes. Death in bacteria and viruses is a retrospective diagnosis. A bacterium is defined as dead when it cannot be grown. If an organism fails to grow when cultured in a broth or on a plate, having previously been successfully cultured, then we can say it is dead. Or we can say that we failed to grow the organism in the correct conditions. Because it is impossible to prove a negative event (absence of growth) there is always the worry that we have not killed the organism but simply failed to grow it. The reasons for organisms *not* growing in laboratory culture media are considerable. If incubated in broth for longer, using a different temperature or different culture broth, we might observe growth and the original conclusion that the organisms was dead would be mistaken.

Of more practical concern to microbiology laboratories is the need to sterilise items. From sterilisation of culture media, necessary for the culture of microbes in clinical samples and their identification to the disposal of clinical samples, sterilisation is a prerequisite and demonstrates how microbiologists need to think in terms of sterility throughout the working day.

## ■ 5.2 KILLING IS EXPONENTIAL

A large number of compounds including chemical disinfectants and antibiotics have been identified that can be used in controlling the multiplication of bacteria. Although the mode of action will differ between different compounds, it is helpful to distinguish between whether they act by killing bacteria, in which case they are termed **bactericidal** or only inhibit bacterial proliferation rather than kill the organism, when they are described as **bacteriostatic** (note the different spelling construction).

Figure 5.1 illustrates the growth curve of bacterial cultures in which bactericidal and bacteriostatic compounds have been added whilst the organisms are in exponential growth. The bactericidal agent reduces the numbers of organisms, while the bacteriostatic compound only prevents any further replication but does not kill the organisms already present. Whilst it is understandable that bacteriostatic compounds are useful in circumstances such as preserving foodstuffs and preventing bacterial multiplication on the butcher's workbench, bacteriostatic agents are inappropriate when attempting to create a sterile medical product. Figure 5.1 also illustrates how the addition of the bacteriostatic compound still has a dramatic effect simply by preventing the exponential increase in bacterial numbers. The difference in the slope of the curve demonstrates how useful a bacteriostatic compound is. In human infections, bacteriostatic antibiotics prevent the increase in bacterial numbers and allow the immune defences to gain control. The neutrophils will be able to engulf and destroy the bacteria that are held in check by the bacteriostatic compound, whereas without this assistance the bacterial numbers may overcome the ability of the neutrophils to cope.

If you count the organisms in a culture exposed to a bactericidal agent (heat or chemical biocide) over a period of time, the numbers will decline progressively rather than instantaneously. Table 5.1 shows the numbers of *Esch. coli* obtained from such an experiment to study the rate of killing when the culture is heated to 65°C. The results are shown in Figure 5.2 and illustrate how misleading it is to plot the numbers of organisms in an arithmetic graph (Figure 5.2a). The slope of the line appears to be hori-

• **Figure 5.1**  Effect of bactericidal and bacteriostatic agents (added at the time point indicated by the arrow) on the numbers of viable organisms in a bacterial culture. The numbers of bacteria are plotted on a linear scale

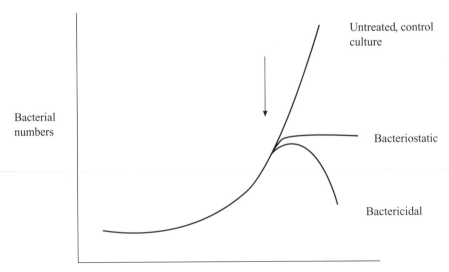

**Table 5.1  Data for survivor curve of *Esch. coli* held at 65°C**

| Time (mins) | Number of organisms | % of survivors killed each time interval | Cumulative % killed |
|---|---|---|---|
| 0 | $10^3$ | 0 | 0 |
| 10 | $10^2$ | 90 | 90 |
| 20 | $10^1$ | 90 | 99 |
| 30 | $10^0$ | 90 | 99.9 |
| 40 | $10^{-1}$ | 90 | 99.99 |
| 50 | $10^{-2}$ | 90 | 99.999 |
| 60 | $10^{-3}$ | 90 | 99.9999 |
| 70 | $10^{-4}$ | 90 | 99.99999 |
| 80 | $10^{-5}$ | 90 | 99.999999 |

zontal from about 20–30 minutes whereas the values in Table 5.1 and Figure 5.2b clearly demonstrate that the slope is the same throughout the 80 minutes. In other words the rate of killing is constant and does not slow down once the numbers get small. When exposed to a lethal agent not all of the bacteria in the culture die immediately but, instead, a proportion of the total will be killed per unit time. As the time course of killing is plotted with a logarithmic Y axis, a straight line is obtained, as shown in Figure 5.2b and thus, somewhat unsurprisingly, is described as **logarithmic**, although 'exponential' is more accurate. (You will notice that this pattern is the same as the description of bacterial growth curve except the sign of the constant is now negative.) The straight line obtained when the data are plotted on semilogarithmic paper tells us that the relationship is a first order reaction; that is, the killing is a constant feature over time and directly proportional to the number of organisms present. This means that the fraction killed will be the same per successive unit of time. Note that this is not the same as the same number of organisms killed per unit time. Instead, the **exponential** nature of the killing means that the absolute numbers of organisms killed is proportional to the number present at that time point. In other words, the more organisms there are, the more organisms are killed.

The rate of killing is the slope of the line and in this case will be negative because we are describing the reduction in bacterial numbers. This relationship between numbers of surviving bacteria and time can be described by the mathematical equation:

$$N/N_0 = e^{-kt} \qquad (5.1)$$

where $N/N_0$ is the number of survivors (number at any particular point in time divided by the number at the beginning ($N_0$)); $e$ is the natural logarithm; $t$ is time and $k$ is the constant that describes the relationship between time and bacterial numbers.

Equation 5.2 employs natural logarithms to show that the rate of killing is represented by the slope of the graph. We can rearrange equation 5.1:

$$(N/N_0) = e^{-kt}$$

by taking natural logs of both sides:

$$\ln (N/N_0) = -kt$$

The problem with an exponential decrease in numbers is that the proportional decrease in numbers means that you cannot obtain zero but simply a smaller fraction of the previous value.

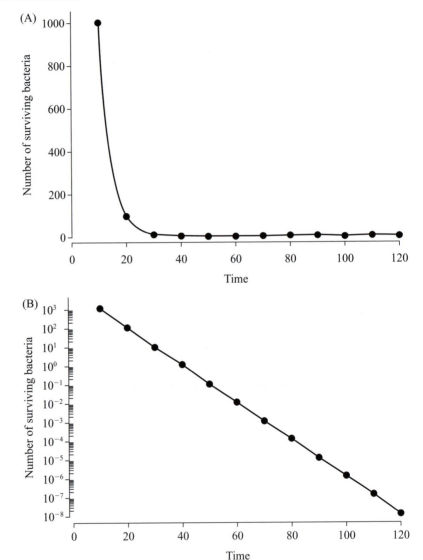

• **Figure 5.2** A survivor curve. The numbers of viable organisms are plotted against time. In graph (A) the *y* axis is linear, whereas in (B) the numbers of bacteria are expressed as logarithms (base 10). Note that the first two points on the graph represent a 90 per cent reduction in bacterial numbers

which, rearranged, gives:

$$\ln N - \ln N_0 = -kt$$

and further re-arrangement describes the form of a straight line graph ($y = ax + b$):

$$\ln N = -kt + \ln N_0 \qquad (5.2)$$

where the inactivation constant ($k$) is the slope of the line, $t$ is the value at any point in time and $N_0$ is the number of organisms you started with and, hence, intercepts the *y* axis at time zero.

Equation 5.2 can be expressed using logarithms to the base 10 because of the following relationship between logarithms of the different bases:

$$\ln x = 2.303 \log_{10} x$$

It is therefore possible to represent equation 5.2 as follows:

$$2.303 \log_{10}(N/N_0) = -kt$$

which can be rearranged to:

$$2.303 (\log_{10} N - \log_{10} N_0) = -kt$$

$$\log_{10} N - \log_{10} N_0 = -kt/2.303$$

and finally:

$$\log_{10} N = -k/2.303 \cdot t + \log_{10} N_0 \qquad (5.3)$$

The value of calculating the **inactivation constant** (**k**) is that it provides a numerical value that can be used in comparisons of killing rates between different temperatures or between different organisms. As the time interval in equation 5.2 is not defined, $k$ could be misleading because it was calculated from a small time interval. In order to make more standardised comparisons an alternative parameter, the $D$ value (decimal reduction time), is widely used in place of $k$. The $D$ value is the time taken for the numbers to fall tenfold (or, in other words, by 90 per cent).

The inactivation constant, $k$ (also known as the exponential death rate constant), has the units of reciprocal time ($t^{-1}$). This can be seen by rearranging equation 5.2:

$$\ln N = -kt + \ln N_0$$

$$\ln N + kt = \ln N_0$$

$$kt = \ln N_0 - \ln N$$

$$k = \frac{\ln N_0 - \ln N}{t} \qquad (5.4)$$

$D$ is a time period and has units of time (usually minutes). $D$ will be the inverse of $k$, and due to the conversion from natural logarithms to logarithms of base 10,

$$D = 2.303/k$$

The lines obtained in killing curves are not always perfectly linear (see Figure 5.3). The reasons are not completely understood but reflect the difficulties in evaluating death in single bacterial cells. Several explanations have been proposed to account for the deviation from logarithmic death:

- delayed killing (shoulders) may represent poor penetration of the lethal agent into clumps of bacteria.

• **Figure 5.3** Patterns of results frequently obtained from killing curves (survivor curves) of bacterial populations. Note how the straight lines have curved shoulders or tails

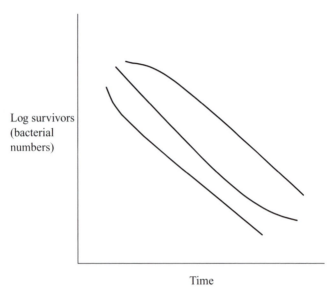

Log survivors (bacterial numbers)

Time

- the lethal event is the sum of several distinct steps (multi-hit hypothesis) rather than a first order reaction (single hit).
- the tails may represent a protective effect of dead cells.
- the organisms are not a homogeneous population. (Straight line survivor curves are often obtained from spore suspensions, which are likely to be less heterogeneous than a population of vegetative cells.)

Leaving our experimental culture plates on the bench in the sun until all the water has evaporated from the agar is one way to kill the microbes (and your research project), but it is almost completely without control or understanding. What controlled means are available to kill organisms? We will consider three approaches:

- **sterilisation,**
- **disinfection,**
- **antimicrobial agents (antibiotics).**

This disparate group of headings is pragmatic and does not reflect any natural order. Using antibiotics is a form of chemical disinfection, although particularly efficacious antibiotics may indeed sterilise a culture. It is important to remember that micro-organisms vary widely in their susceptibility to killing. This variation will be seen both between different species of the same genus and different strains of the same species. Inherent properties of the different genera will account for the large differences in resistance to heat or drying (for example, Gram positive bacteria resist drying better than Gram negative bacteria), whereas smaller differences may occur depending on the environmental conditions. Two strains of the same species may show differences in survival to irradiation depending on the pH of the culture medium or their stage in the growth cycle. Before meaningful comparisons of killing between different organisms can be made, rigorous attention must be given to reduce any variations in culture conditions.

# ■ 5.3 STERILISATION

**Sterilisation** is a process whereby an item is freed of any living micro-organism or their dormant stage. In theory sterility is an absolute state (despite mathematical predictions), hence it is nonsense to use phrases such as 'nearly sterile' or 'almost sterile'.

If we had a large volume of fluid that we wanted to sterilise, the point of absolute sterility can never be obtained because the numbers of organisms remaining continues to reduce by the same fraction per unit time. You will appreciate that this means the number never reaches zero, and absolute sterility cannot be achieved. This may seem a pedantic mathematical artefact but it has implications when you wish to sterilise large volumes. Take, for example, a pharmaceutical company that manufactures 100 ml volume bags of saline solutions for intravenous injection. What duration of time should the bags be sterilised for? The saline solutions start out with $10^3$ organisms in 100 ml it takes 40 minutes to reduce the numbers of organisms to less than one per bag. To be exact, 0.1 organisms per 100 ml. This means that in every 10 bags there will be one bag that contains one organism. If they sterilised 100 bags then potentially there will be 10 bags with a surviving organism present. You would not take the chance that, if you were the patient needing a saline drip, the bag you were given was not sterile. To avoid this problem manufacturers of 'sterile' medical products set the sterilisation time such that the probability of finding one organism alive is 1 in a million ($10^{-6}$). This is called the 'sterility assurance limit'. The time taken to reach 1 in a million is extrapolated from a survivor graph of bacterial spores (being heat resistant, the use of spores creates a worst case scenario) as illustrated in Figure 5.4. In practice, the solutions have less than one spore per ml at the outset so that, after sterilisation, the chances of a product being non-sterile are vanishingly small.

There are different strategies for sterilising items, each having advantages and disadvantages. The methods are: the use of moist or dry heat, filtration, chemical treatment or irradiation.

• **Figure 5.4** *D* values and sterility assurance. The *D* value is the time taken for the numbers of organisms to reduce tenfold. For the organism shown here, the *D* value will be the time difference (usually in minutes) between point A and point A'. The sterility assurance value will be the time taken (from the start) to reach point SA. This represents the time taken for one organism to have been reduced a further six logarithmic reductions

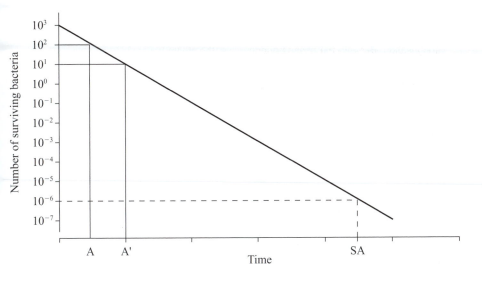

### ■ 5.3.1 MOIST HEAT

Most non-sporing bacteria (i.e. the vegetative form) are killed when heated to 60°C. Yeasts and fungi need temperatures over 80°C. Bacterial endospores, however, are only killed to any significant extent when held at temperatures above 100°C for over 5 minutes. Bacterial endospores, therefore, pose the greatest problems in obtaining sterility. Because water acts as a better conductor of heat than air, the transfer of energy into the microbe is achieved more efficiently when the organisms are heated in moist or wet conditions (i.e. in suspension) rather than on dry material. The resistance of microbes to 100°C as boiling water or dry heat differs considerably.

In order to obtain temperatures above 100°C (i.e. raise the boiling point) of water it is necessary to heat water under pressure. This process is achieved in a pressure cooker at home and, when constructed for sterilisation purposes, is called an **autoclave**. Heating water under pressure raises the boiling point and the released steam is able to transfer the heat (**latent heat of vapourisation**) rapidly into any colder surface material, which in this case will be the micro-organism as the steam condenses. This will then draw more steam into the target (see Box 5.1).

The relationship between temperature and pressure is shown in Table 5.2. The temperatures most commonly employed to sterilise culture media or sterilise biological waste are 115°C, 121°C or 126°C.

At first glance it would seem that any increase in temperature is going to have only beneficial effects on the efficacy of sterilisation. After all, the higher the temperature the greater the rate of killing of bacteria. However, if moisture is driven out of steam you are left with hot air (Box 5.1) and are now simply using dry heat, which is much less efficient at killing micro-organisms, as discussed below.

In practice the biggest problem of autoclaving samples is the complete removal of air. Where air remains, the temperature will be prevented from rising to that obtained

---

### ■ BOX 5.1 PROPERTIES OF STEAM

Steam is the gaseous form of water released once the boiling point is reached. The steam that is required for sterilisation is dry saturated steam. It has no water droplets contained within it but is the same temperature as the water from which it is driven. Above this equilibrium point, also termed a 'phase boundary', the steam becomes **superheated steam**, a 'dehydrated' form of steam that is no longer saturated with water vapour. The steam will need to cool to the appropriate temperature before it can release the latent heat of vapourisation. Beneath the phase boundary line, the steam becomes 'wet' steam. This is too clogged up with water droplets rather than gaseous steam and consequently has lower latent heat energy and the transfer of energy is reduced.

---

Table 5.2  Relationship between temperature of steam under differing pressures

| Temperature (°C) | psi* | kPascals |
| --- | --- | --- |
| 115 | 10 | 69 |
| 121 | 15 | 103 |
| 126 | 20 | 138 |
| 134 | 30 | 207 |

* psi: pounds per square inch.
Pressures given are above standard atmospheric pressure.

by steam under pressure (Figure 5.5). Also, the penetration of steam into objects is not always guaranteed. A closed container such as a bottle or closed bag will not permit entry of the steam and remain 'dry'. Again, dry heat is far less efficient in its killing potential.

Autoclaves are usually employed to sterilise items that are not adversely affected by moisture, such as culture media. The sterilised items will emerge from autoclaving wet. This is a problem if the items to be sterilised are powders or woollen blankets!

## ■ 5.3.2 DRY HEAT

Sterilisation by dry heat takes higher temperatures for longer periods than wet heat. Steam at 121°C will sterilise a spore suspension in 15 minutes, whereas at 120°C in dry heat you will need 8 hours! For this reason hot-air ovens are usually held at 160–180°C for 2 hours.

Heat can be applied to a material in various forms. Incineration or passing through a Bunsen flame are two forms of applying dry heat, but usually the end product is not in much of a fit state to use. Any means of holding the items at a raised temperature for long enough will coagulate proteins in the microbes but the risks of damaging the item itself usually limit the range of applications of dry heat sterilisation. Hot-air ovens are used to sterilise products that must not get wet and are heat stable. These include powders, certain surgical instruments and glassware. The transfer of heat is a slow process and insulated and/or large objects that heat slowly need to be accommodated for in the duration of the heat cycle.

### 5.3.2.1 Controls

In the use of autoclaves and hot-air ovens it is essential that the process is monitored. The temperatures quoted are never obtained instantaneously and, likewise, glassware in hot-air ovens will take a long time to cool. Hence both operate via automated

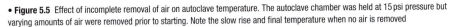

• **Figure 5.5** Effect of incomplete removal of air on autoclave temperature. The autoclave chamber was held at 15 psi pressure but varying amounts of air were removed prior to starting. Note the slow rise and final temperature when no air is removed

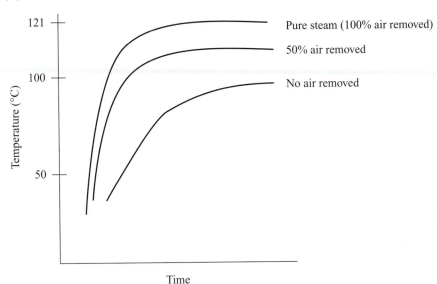

control panels that can be programmed to take into account the heating up and the cooling down phases of the cycle and, most importantly, the holding time, the period at which the load is kept at the required temperature. Thermocouples are placed inside mock large objects that represent the point of least penetration by the steam. Although measurements of the temperature are important, biological controls offer a direct measure of the sterilising power of the system. For this purpose suspensions of bacterial spores (commonly *Bacillus stereothermophilus* spores) are placed inside the autoclave and then tested for growth once the cycle is finished. This will measure the effectiveness of the sterilisation, albeit taking a further 24–48 hours and rigorous aseptic technique. Chemical dyes, known as Brownes tubes, can also be used. These change colour once a certain temperature has been reached but give no idea of the length of time the temperature was maintained.

### ■ 5.3.3 IRRADIATION

The waves of energy carried by electromagnetic radiation can kill micro-organisms by a number of mechanisms, depending on their frequency. Ranked from lowest to highest frequency, the different types of radiation that are useful in sterilisation are infrared, ultraviolet, X-ray and γ-rays.

#### 5.3.3.1 Infrared radiation

Infrared radiation raises the temperature of objects rapidly (like microwaves!) so that its sterilising ability acts as dry heat sterilisation. Under vacuum the temperatures rise further.

#### 5.3.3.2 Ultraviolet radiation

UV radiation ranges between 210–330 nm, but those around 260 nm are efficiently absorbed by nucleic acid and cause dimer formation between adjacent thymines, with lethal, mutagenic effects. The biggest drawback of UV light is the very poor penetrative power, such that it is ineffective through thin layers of dust or glass.

#### 5.3.3.3 Ionising radiation

Ionising radiation such as X-rays and γ-rays are of short wavelength (<1 nm) and possess much higher energy and greater penetrating power than non-ionising wavelengths. As the rays pass through the cell they remove electrons from molecules thus creating ionised (charged) molecules. When water is ionised it forms hydroxyl radicals that initiate destructive free radical chain reactions which are damaging to most nucleic acid and proteins.

Sterilisation by irradiation is not readily available in laboratories but will be used by the pharmaceutical industry to sterilise single-use medical devices.

### ■ 5.3.4 FILTRATION

Temperature-sensitive substances such as carbohydrates and amino acids that will be destroyed by heat sterilisation need to be sterilised by filtration. The pore size (0.2 μm pore diameter) will prevent bacteria passing through the filter. Filtration does not involve killing organisms but rather simply trapping them on a filter, small enough in effective 'pore' size that the organisms do not pass. Whilst filtration has no destructive effect on organisms or the product, the released products such as Gram negative cell wall (LPS) or even secreted toxins will not be retained by filtration. This is a potential risk in solutions for medical administration such as saline drip solutions. Similarly, viruses will not be retained by filters of 0.2 μm pore size. Special ultrafiltration membranes are needed to filter viruses.

### ■ 5.3.5 CHEMICAL STERILISATION

Most chemicals cannot be relied upon to sterilise items, but rather act as disinfectants. There are two chemical compounds that are employed as sterilising agents: ethylene oxide and formaldehyde.

**Ethylene oxide** ($(CH_2)_2O$) is a colourless gas that when used as a gas is able to kill both vegetative cells and spores. It is toxic and potentially inflammable but is used in pharmaceutical and veterinary applications.

**Formaldehyde** is a toxic gas (formalin is the commercial solution of 40 per cent formaldehyde) which whilst able to sterilise exhaust cabinets and rooms at low concentrations (1 per cent) should not be inhaled, for obvious reasons. Formaldehyde is also used in combination with low temperature steam sterilisers for sterilising heat-labile items such as medical instruments (e.g. fibre-optic bronchoscopes). The heat-sensitive instruments are kept at 73°C for 2 hours and *Mycobacterium tuberculosis* will be killed by this procedure. The related aldehyde, glutaraldehyde, is also an effective sterilising solution.

In their favour, chemical methods of sterilisation overcome the limitations posed by heat-labile items but unfortunately leave a potentially toxic chemical residue.

### ■ 5.4 DISINFECTION

Disinfection is defined as the destruction or removal of organisms capable of giving rise to infection. This partial selectivity means that disinfection reduces microbial numbers to below the infectious dose and does *not* imply sterility. Disinfection is often a description of the use of chemical agents but physical methods such as heat treatments are also included. Where sterilisation is impossible (e.g. human skin) or practically unrealistic (sterilising toilets and hand basins) disinfection is of great practical importance. Care and thought is needed to obtain successful results.

**Antiseptics** are chemical disinfectants that have suitably low toxicity such that they can be applied to mammalian tissues in order to prevent infection.

Disinfection need not mean the addition of toxic chemicals! Strictly, disinfection will include heating or simple washing with detergent. Both procedures will remove significant numbers of organisms so that the risk of infection is virtually eliminated. Nevertheless, disinfection with chemical disinfectants means that you will also kill a proportion of the organisms irrespective of whether you remove them in the bucket along with the soap and water. And, to be exact, soaps will also inhibit organisms.

Figure 5.6 presents a diagram representing the route by which organisms are

As a general rule chemical disinfectants work well and are used at high concentrations but are toxic, whereas antiseptics are less efficacious, used at low concentrations and are non-toxic.

• **Figure 5.6** Route of transmission. An infection can be viewed as the movement of an organism through four stages. Disinfection aims to interrupt the flow between stages 2 to 3

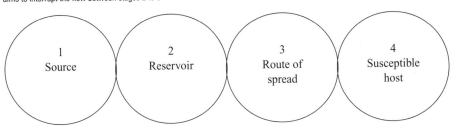

**Table 5.3 Chemical disinfectants**

| Group | Example | Concentration | Inactive against |
|---|---|---|---|
| Alcohols | isopropanol | 90% v/v | spores/viruses |
| Halogens | iodine | 2% w/v | |
| | chlorine | 1ppm | |
| Phenolics | cresol, hycolin | 0.5–3% v/v | |
| Biguanides | chlorhexidine | 1–4% v/v | spores/ mycobacteria |
| Quaternary ammonium compounds (QACs) | cetrimide | 0.1% | |

transferred in the sequence of events that result in an infection. The usual sequence of events follows the diagram from left to right. Disinfection aims to break the connection between the reservoir and the route of spread. In the context of a hospital ward the source of a micro-organism is usually unknown, but application of antibacterial hand cream following hand washing by the hospital will help minimise the risk of transfer of organisms from the wound of patient X (reservoir) to the next patient (susceptible host). The concept of reservoir and source will be examined further in Chapter 7.

### ■ 5.4.1 CHEMICAL DISINFECTANTS

There are a large number of chemicals that exert antimicrobial activity. Unlike antibiotics, they need relatively high concentrations to work and all are inactivated by dirt or organic matter. The target sites are often not clearly defined but include disrupting cell wall and cell membrane integrity or oxidising/denaturing proteins within the cytoplasm.

The major groups of disinfectants that are used in hospital and laboratory settings are given in Table 5.3. There are two particular challenges to the use of disinfectants in killing microbes. These are the i) variation in susceptibility of different micro-organisms to disinfectants, and ii) the inactivation of disinfectants in their working environment.

(i) Perhaps not surprisingly, the diversity in structure of micro-organisms manifests itself in a spectrum of tolerance to disinfectants, ranging from susceptible through to resistant. The sequence of decreasing resistance is shown in Figure 5.7.

(ii) The environment in which the disinfectant is expected to work can vary enormously and often has a number of features that inactivate or impair their bactericidal action. One of the most problematic features is the presence of organic matter. Organic matter will compete with the microbes as the target for the disinfectants and either exhaust or neutralise the active component of the disinfectant. The problem is easily demonstrated by considering what to do about disinfecting any body fluid spill (sputum, urine, blood) onto a floor or uniform. The organic nature of the spill poses a serious challenge to any effective killing by disinfection of any microbes present with this predominantly organic cocktail.

Before addressing this problem, other factors that alter the efficacy of disinfectant action need to be discussed:

• **Figure 5.7** Spectrum of resistance of different micro-organisms to chemical disinfectants

**MOST RESISTANT**

Mycobacteria

Bacterial spores

Non-enveloped viruses

Fungi (yeasts and moulds)

Vegetative forms of bacteria

Enveloped viruses

**LEAST RESISTANT**

- duration of exposure. It appears all too easy to assume that having sprayed the worksurface with a disinfectant (e.g. 70 per cent alcohol) and then wiped it down with paper cloth the surface is suitably disinfected. Disinfectants do not achieve their desired effect instantaneously but need time to act.
- population size. As discussed earlier, the concept of exponential killing dictates that the larger the initial population of microbes the longer it will take for the number to be reduced to an acceptable level.
- disinfectant concentration. Within limits, an increase in concentration will hasten the disinfection process.
- temperature. Raising the temperature will also hasten the reduction in microbial numbers as the speed of reaction increases. In fact, the relationship between the two is not linear. The rate of microbial killing goes up geometrically with linear increases in temperature. Clearly, extremes of temperature may confound the disinfection process by either inactivating the chemicals themselves or promoting the growth of the microbes.

> The guiding principle for use of disinfectants is 'strong enough for long enough'.

Is there then any hope for disinfectants achieving any significant reduction in the microbial load? Disinfectants are prepared at recommended concentrations following testing under experimental conditions in which the above variables are examined. In hospital wards and laboratories recommended procedures will be in place which should provide a level of protection that is acceptable for most encounters. Fortunately, the risk of infection from spills is usually low due to the unlikely chances of an infectious dose being transmitted in an appropriate manner.

If, by cleaning with warm soapy water, you remove most interfering organic matter along with a proportion of the microbes why use disinfectants? Disinfection of surfaces like walls and floors or equipment that pose negligible risk of infection to staff will be a waste of expensive disinfectant. Under such circumstances the question is entirely

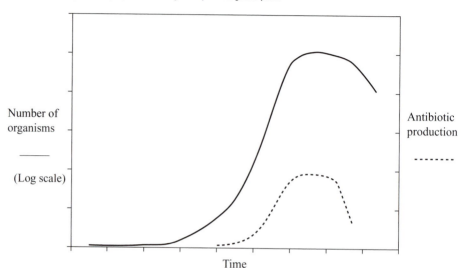

• **Figure 5.8** Antibiotic production (typical of secondary metabolites) is shown in relation to the growth phase of the organism (solid line). The antibiotic (dotted line) is produced during late exponential growth phase

which are essential for their own growth)) and compounds that act as metabolic ana-logues (i.e. compounds that resemble normal metabolites or substrates but are antago-nistic to growth through inhibition). The latter are termed 'antimetabolites' or 'chemotherapeutic agents' rather than antibiotics. Metabolic analogues are chemically synthesised and are not produced by micro-organisms.

The number of different organisms that the antibiotic works against is called the **spectrum of activity**. This varies for different antibiotics, with some having a broad spectrum of activity (many different organisms inhibited) whereas others have a nar-rower range of action. The spectrum of activity can be altered by chemical modification of the antibiotics, notably the penicillins.

The ideal features of an antibiotic for human use are as follows:

(i) **selective toxicity.** This term describes a property of the antibiotic in which the target site is present in the bacterium but not the host. Selective toxicity can be expressed numerically as the selective index, the ratio of 50 per cent value of toxic dose to 50 per cent inhibitory dose ($TD_{50}/IC_{50}$). When assessed *in vivo* the therapeu-tic index can be assessed as the ratio of toxic dose to the effective therapeutic dose.

For peptidoglycan in bacterial cell walls, the selective toxicity is obvious as humans lack such a structure. However, many antibiotics rely on more subtle dif-ferences. For example, the differences in ribosomal subunit structure between prokaryotes and eukaryotes. The targets consequently become more problematic still when targeting fungi and viruses.

(ii) **appropriate pharmacology.** Does the antibiotic reach the site of infection? Having found a suitable target with your new antibiotic, it is essential that the phar-macological properties are suitable for use in patients if there is any hope of bring-ing the compound to market. Variation in pharmacokinetics are useful. Some antibiotics are not absorbed across the intestinal tract and need to be injected intra-venously. Then you might find that the drug concentrates in liver but you wish to treat a urinary tract infection and need an antibiotic that is excreted via the kidney.

(iii) **retain efficacy.** The antibiotic needs to avoid inactivation by neutralising agents (if urea in urine interferes with the antibiotic then treatment of urinary tract infection will be compromised). Conversely, the route of elimination of the antibiotic from the body can help direct the drug to the appropriate site (e.g. renal excretion is suited to the treatment of urinary tract infection) but many antibiotics are eliminated from the body very quickly, necessitating repeated dosages.

## ■ 5.7 ANTIBIOTIC TARGETS

The variety of target sites exploited for developing drugs for use in treating human infections are given for antibacterial, antiviral and antifungal compounds. The antibiotics discussed are chosen on the basis of their selective toxicity and are summarised in Table 5.5. In other words, the important points to consider are the targets within the microbes that are sufficiently different from mammalian cells to minimise toxicity to humans but still inhibit the microbes. The events that actually kill the bacterium are surprisingly hazy in most cases despite apparent understanding of the molecular targets. Appropriate texts on antibiotic action should be consulted for details.

## ■ 5.7.1 ANTIBACTERIAL COMPOUNDS

The target sites of greatest value to date are as follows:

- bacterial cell wall,
- cytoplasmic membrane,
- protein synthesis,
- nucleic acid synthesis and replication.

### 5.7.1.1 Bacterial cell wall

Peptidoglycan provides the bacterium with a rigid cell wall to confer strength and resistance to turgor pressure from the cytosol. **β-lactam** antibiotics (penicillins and cephalosporins) target the assembly of the peptidoglycan. The polymers of NAG-NAM molecules are cross-linked by peptide chains. The controlled insertion of new

**Table 5.5** Antibiotic targets and mode of action

| Structure | Target | Effect |
|---|---|---|
| | CELL WALL | |
| β-lactams | Penicillin binding proteins | Impair cell wall synthesis, trigger autolysis |
| | PROTEIN SYNTHESIS | |
| Aminoglycosides, e.g. Streptomycin | 30S ribosomal subunit | Misreading |
| Tetracycline | 30S | Inhibit tRNA |
| Erythromycin | 50S | Inhibit translocation |
| Chloramphenicol | 50S | Inhibit peptidyl transferase, thus transpeptidation |
| | NUCLEIC ACID | |
| Fluoroquinolones | DNA gyrase | Inhibit DNA replication |
| Sulphonamides | Dihidropteroate synthetase | Inhibit folate synthesis |
| Trimethoprim | Dihydrofolate reductase | Inhibit thymidine synthesis |

NAG-NAM units is essential if the peptidoglycan integrity is not to be compromised. There are several different types of enzymes that facilitate this process. Functionally, the enzymes are transpeptidases and carboxypeptidases, collectively called **penicillin-binding proteins** (PBPs) as they are the targets for the β-lactams. There are between five and seven different PBPs per species and different PBPs bind different β-lactams. The β-lactams impair the extension of the cell wall as the organism attempts to grow. The exact reasons for the cell death induced by β-lactam antibiotics is not clear but the triggering of bacterial autolysins is suspected. Such enzymes may be PBPs themselves in that they regulate the transient disassembly of the peptidoglycan whilst new peptidoglycan is inserted. Perhaps when bound by penicillins they are irreversible in their action. Unrestricted, the osmotic pressure will burst the cytoplasmic membrane if the rigid peptidoglycan is breached. Certainly, β-lactam-treated bacteria are able to survive if they are kept in osmotically balanced solutions the bacteria form **spheroplasts** and **protoplasts**, viable bacterial forms that have only a fraction of the normal complement of cell wall (spheroplasts) or no cell wall at all (protoplasts).

The location of the PBPs in the bacterium has significant impact on the efficacy of β-lactam antibiotics between Gram positive and Gram negative bacteria. With Gram positive bacteria the target (peptidoglycan) lies outside the cytoplasmic membrane and the β-lactams, being water soluble, have free access to their targets. Gram negative bacteria will have the outer membrane external to the cytoplasmic membrane and, being a permeability barrier to aqueous molecules, it will limit the rate of entry of β-lactams to the peptidoglycan. The hydrophilic β-lactams pass through the outer membrane via water-filled **porins**.

### 5.7.1.2 Cytoplasmic membrane

Interior to the bacterial peptidoglycan is the cytoplasmic membrane. Of the differences that exist between mammalian and bacterial cytoplasmic membranes, of most relevance is the overall surface charge. Mammalian phospholipid membranes are predominantly zwitterionic (possessing both negative and positive charge and the major component is phosphatidylcholine) whereas prokaryotes contain a net negative charge through the dominance of anionic phospholipids. One antibiotic that preferentially binds to negatively charged membranes is **polymixin**, representative of the class of **cyclic peptide** antibiotics. Being polycationic, polymixin will bind to negatively charged phospholipids and disrupt the membrane integrity with subsequent osmotic perturbation (leakage of intracellular contents). You will recall that the outer leaflet of the outer membrane of Gram negative bacteria does not contain phospholipid, but, negatively charged lipopolysaccharide (LPS). Polymixin will also bind to LPS but this binding will not result in disruption of the cytoplasmic membrane.

A similar affinity for negatively charged phospholipids underlies the action of epithelial **antibacterial peptides**, compounds secreted by mammalian epithelial cells in response to microbial infection. These peptides form part of the innate response to infection and are expressed in plants, and insects as well as animals, and are under investigation for use as therapeutic agents. The peptides (and there are a growing number identified) are cationic and small (less than 100 amino acids) and have a broad spectrum of antibacterial action. The peptides bind to and disrupt both the LPS and cytoplasmic membranes. The selective toxicity lies both in the affinity for anionic bacterial membranes and because they are inhibited by the cholesterol present in mammalian membranes. Whilst not true antibiotics, they are hoped to provide a new line of treatment. Optimism might be considered justified because these peptides have a number of advantages: selective toxicity, broad spectrum of activity, rapid killing time through acting at more than one site. The sites and mechanisms of action mean that the bacteria cannot develop resistance very readily.

### 5.7.1.3 Protein synthesis

A number of antibiotics act by inhibiting aspects of protein synthesis (chloramphenicol, erythromycin, tetracycline and the aminoglycoside group of antibiotics). Protein synthesis occurs at the ribosome and selective toxicity arises from differences between eukaryotic and prokaryotic protein at the ribosome. This is reflected in the fact that eukaryotic ribosomes are larger than those in prokaryotes (see Figure 1.13). These differences have a number of implications for their cell biology.

*Size of ribosome*

- Considerably more associated proteins and cofactors are required for eukaryotic protein synthesis (e.g. approximately 10 initiation factors compared with 3 in bacteria). In terms of making the ribosome and associated factors, there are metabolic savings to be made if fewer proteins are required and the size of the ribosome is smaller.

*Speed*

- Protein synthesis is faster in many bacteria than eukaryotes (e.g. humans). The spatial restriction of the mammalian genome inside the nuclear membrane means that mRNA travels further to reach the cytosolic ribosome. No such nuclear membrane exists in prokaryotes, hence turnover is approximately tenfold greater.
- Bacterial mRNA is polycistronic, i.e. numerous different proteins are coded for on the same mRNA (often as an operon).

*Output*

- mRNA is translated by several ribosomes simultaneously (polysome), thereby increasing the rate of production of proteins.

Examples of antibiotics that target protein synthesis (see Figure 5.9) are as follows.

### ■ 5.7.2 ANTIBIOTICS THAT TARGET THE 30S SUBUNIT

#### 5.7.2.1 Aminoglycosides

Streptomycin and gentamicin are the most representative members of this large group of antibiotics. Streptomycin was discovered in the 1940s and became the first antibiotic to be used against *Mycobacterium tuberculosis*. Unfortunately, resistance readily develops against streptomycin and it is nephrotoxic and ototoxic. Gentamicin is used for serious infections and is preferred because of its broad antibacterial spectrum (which is considered to be bactericidal rather than just bacteriostatic), although it is still toxic.

The reason why aminoglycosides are bactericidal remains unclear despite considerable research. Aminoglycosides have two effects: the misreading of the mRNA chain and prevention of initiation of protein synthesis. These two events may appear contradictory. How do you cause misreading if you have prevented the process from starting? It is possible that the two events are induced by differing concentrations of antibiotic.

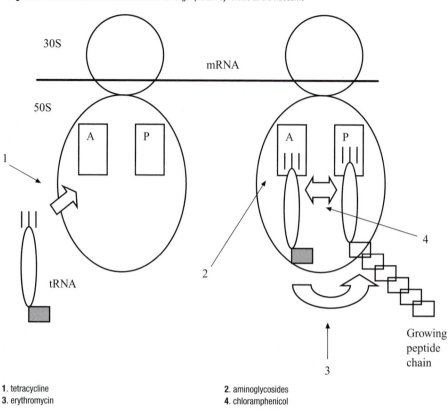

• **Figure 5.9** Sites of action for the antibiotics that target protein synthesis at the ribosome

**1**. tetracycline
**3**. erythromycin
**2**. aminoglycosides
**4**. chloramphenicol

At low concentrations, streptomycin causes misreading of pre-existing and active processing in the abundant polysomes; but it is only at higher concentrations that the antibiotic is able to prevent the initiation of new ribosomal activity.

### 5.7.2.2 Tetracyclines
(Tetracycline, minocycline, doxycycline) are broad spectrum antibiotics. Resistance has developed and they are less used than previously, although newer derivatives have been developed. Tetracyclines block the binding of tRNA to the A (acceptor) site on the 30S subunit.

### ■ 5.7.3 ANTIBIOTICS THAT TARGET THE 50S SUBUNIT

### 5.7.3.1 Chloramphenicol
Chloramphenicol is readily absorbed from the gut and distributes itself widely throughout the body. This made it useful for treatment of infections in relatively inaccessible organs such as the central nervous system (meningitis) and the gall bladder (typhoid), but is not now widely used because of bone marrow suppression (aplastic anaemia) and widespread resistance.

Chloramphenicol prevents the transfer of the growing polypeptide chain from the P (peptidyl-tRNA) site to the next aminoacid-tRNA in the A site. The enzyme that catalyses the transfer is called peptidyl transferase and it is this process (transpeptidation) that is inhibited. Note that the peptide chain is the effective target rather than the ribosome.

## 5.7.3.2 Erythromycin

This antibiotic represents the class called macrolides, which have a broad spectrum of activity. Typically erythromycin can be used in patients allergic to penicillins. Erythromycin prevents the translocation of the peptide chain and tRNA (peptidyl-tRNA) from the A to the P site on the 50S subunit.

### ■ 5.7.4 NUCLEIC ACID FUNCTION

There are a number of compounds that bind directly to DNA either as **intercalators** (ethidium bromide), **cross-linking agents** (mitomycin C) and **alkylating agents** (nitrogen mustards). Intercalators insert in between the base pairs in DNA. Usually flat, planar molecules they distort the reading frame such that frame shift mutations occur in the copying of the DNA. One of these, acridine orange, is fluorescent and used to detect bacterial DNA. All three types of compounds are useless as antibiotics as a result of the absence of selective toxicity. More important are the compounds that follow.

### 5.7.4.1 DNA replication

Quinolones (nalidixic acid, ciprofloxacin) inhibit the function of DNA gyrase (DNA topoisomerase II). These enzymes catalyse the unpacking of the supercoiled DNA ready for DNA replication or transcription by RNA polymerase. Once DNA packaging is disrupted, DNA synthesis ceases. You will recall that when unravelled, bacterial DNA is around 1000 times longer than the bacterial cell body. Any interference with such a critical operation makes sense as a good target for antibiotic action. Unfortunately, it is not entirely clear what the lethal events are in the action of quinolones. Selective toxicity derives from differences in subunit composition of the DNA gyrases between man and bacteria.

### 5.7.4.2 DNA synthesis

**Sulphonamides** and **trimethoprim** are not true antibiotics but chemotherapeutic agents as they are synthetically devised and manufactured. They target two distinct steps in the bacterial synthesis of folate, a precursor of the purine base thymidine. Selective toxicity derives from the fact that man cannot manufacture folate but instead absorbs it entirely from the diet, hence humans lack the relevant enzymes that these antibiotics target. Sulphonamides are homologues of para-aminobenzoic acid (PABA) and act as competitive inhibitors, thereby blocking the manufacture of dihydrofolate.

Figure 5.10 shows that trimethoprim acts on an enzyme further down the folate pathway. The significance of this is of some interest. In the laboratory, if the two antibiotics are used together the inhibition of growth is synergistic (the inhibition of the two is greater than the addition of the effect of the individual antibiotics). The reasons for the synergy have been debated for many years. It is possible that this synergy is due to the combined action on two different targets of the same pathway. However, it has been argued that if the pathway is linear, any inhibition of an enzyme at one point in the pathway will be overcome because of the build up of substrate (because of the build up of substrate prior to the step catalysed by the affected enzyme). The synergy occurs only if the pathway is cyclical rather than linear because then all components will be affected when one enzyme in the cycle is inhibited.

### ■ 5.7.5 ANTIVIRAL COMPOUNDS

The exploitation of the mammalian host cell machinery by viruses poses numerous challenges to the development of antiviral compounds that exert a selective inhibition

• **Figure 5.11** Action of acyclovir. Acyclovir acts as a substrate for the viral thymidine kinase (TK). The viral TK, but not human kinases, phosphorylate the acyclovir such that when incorporated into new viral nucleic acid, it prevents DNA replication

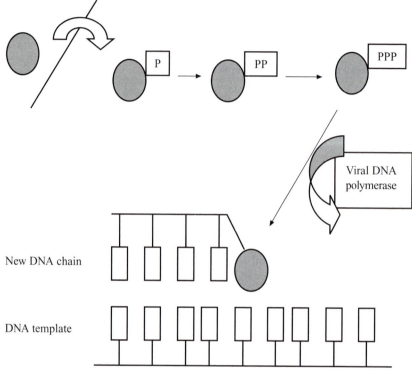

### 5.7.5.2 Reverse transcriptase inhibitors

Reverse transcriptase (RT) catalyses the synthesis of DNA from RNA. The inhibitors of retroviral RT such as AZT are also nucleoside analogues and thus work using the same process as acyclovir, namely inhibition of DNA chain elongation following incorporation into the new chain. Retroviruses do not synthesise their own thymidine kinase so AZT is phosphorylated by the host cell kinases. This means that uninfected cells will metabolise AZT. The selective toxicity has to exploit the unique possession of RT by retroviruses but not eukaryotes. The AZT must inhibit RT but have nominal effect on host DNA polymerase. This is the difference between acyclovir and AZT: acyclovir targets viral TK whereas AZT targets viral RT, although the end result is the same.

### 5.7.5.3 Protease inhibitors

That viruses use host cell machinery for synthesis of viral proteins makes the identification of specific antiviral agents that target protein synthesis very challenging. One key difference that has been identified is the processing of proteins after synthesis (post-translational modifications). Eukaryotic mRNA is monocistronic; that is, it codes for only one protein. Viral mRNA by contrast is polycistronic. The different proteins coded for on the single mRNA are synthesised as a large polyprotein and need to be cleaved into the different proteins by specific, virally encoded proteases. One such protein that has been targeted is the HIV protease that releases viral proteins from the precursor protein, thereby arresting the production of a new virus.

Analysis of the structure has identified potent inhibitors that target the active site of the protease. Unfortunately, the high mutation rate seen in RNA viruses results in mutations in the protease that reduce the affinity of the inhibitors. This has forced the use of multiple protease inhibitors in treatment, reminiscent of the use of multiple antibiotics against single target sites.

These three target sites have yielded only two distinct strategies for antiviral agents. More ambitious strategies to inhibit viral replication have been proposed, namely the use of antisense oligonucleotides. Theoretically, a sequence of nucleotides can be made that will hybridise to viral mRNA and interfere with normal RNA processing. The anti-sense nucleotides need to exhibit increased hybridisation affinity (i.e. raise thermal dissociation, $T_m$) and be chemically modified to reduce degradation by host cell nuclease activity. An alternative new development is the use of **ribozymes**, RNA molecules that hybridise with the target RNA and then catalyse a break in the target RNA strand.

There are other possible targets (release of new virus via budding, for example) that are under investigation, but numerous specificity and drug delivery issues need to be resolved.

## ■ 5.7.6 ANTIFUNGAL COMPOUNDS

Being eukaryotes, fungi possess a number of features in common with mammalian cells and this poses restrictions on target sites when devising a selective antifungal agent. Fortunately a number of suitable targets can be identified that are unique to fungi. Most antibacterial antibiotics do not work against fungi and this highlights the successful targeting between prokaryotic and eukaryotic cell machinery/biochemistry.

The most useful clinical antifungal antibiotics have two targets:

* cell membrane,
* nucleic acid function.

### 5.7.6.1 Nucleic acid function

Chemically, 5-flucytosine is a fluoropyrimidine, i.e. a pyrimidine base with a fluoride group attached. This description indicates that 5-FC works by mimicking nucleotide bases. 5-FC is biologically inert but converted (deaminated) to the 'biologically active' 5-fluorouracil which is incorporated into RNA in place of the normal base uracil. Toxicity correlates with the extent to which 5-FC accumulates in mRNA and tRNA, presumably by interfering with DNA replication (5-FU interferes with thymidine production) and protein synthesis (malfunctioning RNA).

The selective toxicity lies in the fact that most fungi actively transport 5-FC across the cell membrane but mammalian cells cannot as they lack the transporting enzyme (cytosine permease). Cytosine permease transports other nucleosides, adenine and cytosine. There are no transporters for 5-FU in either fungi or mammalian cells. This is important because 5-FU is toxic to mammalian cells (5-FU can be used as a chemotherapeutic agent for treating cancer).

### 5.7.6.2 Cytoplasmic membrane

Two classes of antifungal drugs (**Polyenes** and **Azoles**) exert their effects by interfering with the sterol **ergosterol** present in the cytoplasmic membrane. Fungi use ergosterol whereas mammals use the related compound cholesterol. These membrane steroids act to stabilise the bilayer structure and modulate membrane bound enzymes (Box 5.2). Selective toxicity arises because fungi synthesise ergosterol whereas

---

■ **BOX 5.2 FUNCTION OF ERGOSTEROL IN FUNGAL CYTOPLASMIC MEMBRANE**

- Membrane integrity.
- Membrane fluidity.
- Function of membrane bound enzymes.
- Necessary for progression through cell cycle ('sparking').

---

mammalian cells simply incorporate exogenous cholesterol (from the diet), plus the affinity of drug-binding to ergosterol is higher than for cholesterol. Polyenes (**nystatin** for topical use and **amphotericin B** for systemic use) have a broad spectrum of activity against fungi possibly because they target ergosterol itself rather than its synthesis and ergosterol is found widely in many fungi. The polyenes bind to ergosterol (Figure 5.12) and aggregate to form pores in the membrane of 8 Å in diameter (hydrated sodium ions are 2 Å in diameter). This breach in membrane integrity results in osmotic disruption and leak of cytosolic components (of appropriate size).

**Azoles** are totally synthetic antifungals, classed as imidazoles or triazoles depending

• **Figure 5.12** Polyene action on fungal plasma membranes

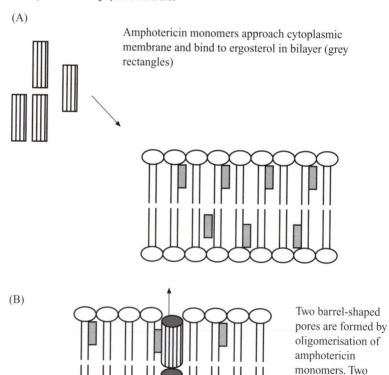

(A)

Amphotericin monomers approach cytoplasmic membrane and bind to ergosterol in bilayer (grey rectangles)

(B)

Two barrel-shaped pores are formed by oligomerisation of amphotericin monomers. Two separate barrels are formed, one on top of the other in order to traverse the entire membrane bilayer on the number of nitrogens in an azole ring present in the structure. The target site is an enzyme involved in the modification (oxidation) of the precursor steroid (lanosterol) earlier in the biosynthetic pathway for ergosterol. The accumulation of precursors of ergosterol cannot replace the functions that ergosterol fungus and the membrane function suffers in terms of loss of membrane integrity and impaired functioning of membrane bound enzymes.

## ■ SUMMARY

Efforts to control bacterial numbers consist of three broad strategies: sterilisation, disinfection and the use of antibiotics. A spectrum of selectivity exists with the first of these processes being completely non-selective in the targets (all organisms) through to antibiotics being used to target single species. Sterilisation is a probability rather than a certainty because of the exponential nature of death in microbial populations. The rate of killing of bacteria can be described as exponential decay in which the numbers of organisms fall proportional to how many are left. If plotted as logarithms, the numbers decrease linearly with time. Sterilisation uses physical (e.g. heat) or chemical (e.g. formaldehyde) agents to kill all living microbes from a material. Filtration may also sterilise a solution by physical removal of micro-organisms.

Disinfection is the destruction or reduction in numbers of organisms that may cause disease. As sterility is a theoretical concept (in that bacterial numbers will never reach zero), the probability of sterility is chosen (sterility assurance) which is considered the best balance between theory and practice. Antibiotics are microbial products that act preferentially on bacterial targets. Their therapeutic value lies in the selective toxicity. The inactivation of a suspension of virus particles is quite straightforward, and greater ingenuity is needed to target viruses within human hosts. As with antifungal agents, the selective toxicity of antiviral agents is of greater importance in order to avoid toxicity to the eukaryotic host.

## RECOMMENDED READING

Challand, R. and Young, R.J. (1997) *Antiviral Chemotherapy*, Spektrum Academic Publishers, Oxford, UK.

Franklin, T.J. and Snow, G.A. (1998) *Biochemistry and Molecular Biology of Antimicrobial Drug Action*, 5th edition, Kluwer Academic Publishers, Dordrecht, The Netherlands.

Hugo, W.B. and Russell, A.D. (1998) *Pharmaceutical Microbiology*, 6th edition, Blackwell Scientific Publications, Oxford, UK.

Russell, A.D., Hugo, W.B. and Ayliffe, G.A.J. (eds) (1998) *Principles and Practice of Disinfection, Preservation and Sterilization*, 3rd edition, Blackwell Scientific Publications, Oxford, UK.

## REVIEW QUESTIONS

*Question 5.1*    What are the proposed explanations for the deviations in linearity for a survivor curve?

*Question 5.2*    What is the basis for selective toxicity for polyenes?

*Question 5.3*    What is the D value for antibacterial agents?

*Question 5.4*    What is the proposed explanation for why wet heat has greater antimicrobial activity than dry heat at equivalent temperatures?

# 6

# MICROBIAL TAXONOMY

The importance of correctly identifying and naming five, ten and twenty pound notes has never been questioned, yet the idea that microbes should be treated with the same attention to detail comes as a surprise to many students. Microbial taxonomy may seem a touch irrelevant to a book on the mechanisms of human infections with micro-organisms, but the elements of taxonomy are critical in understanding the evolutionary development of microbes. In addition, taxonomy is needed to help identify, name and classify organisms that infect humans. Just as any chef needs to know the names and relationships of his foodstuffs in order to work, so microbiologists need to organise rules for systematically organising microbes. With increasing recognition of the value of biodiversity within the planet, microbial taxonomy will play an essential part in defining and documenting huge numbers and types of bacteria that exist as well as prevent confusion in their identification. All students and perhaps especially those of biological and biomedical sciences face an enormous new vocabulary. To memorise the names is important but an understanding of the grammar, that is, the rules governing microbial taxonomy, will be of greater benefit in the long run.

## ■ 6.1 SYSTEMATICS OR TAXONOMY?

'Systematics' is the term used to define the study of the diversity of life and their relationships, whereas 'taxonomy' tends to be restricted to the theory and practice of classifying organisms. Classification attempts to group organisms according to their similarity. In view of the importance of evolutionary pressure on all life, classification of organisms should reflect their evolutionary history. The big questions concerning the evolutionary history of all life forms are being addressed through a form of taxonomy called **phylogenetic systematics**. Phylogeny investigates the relationships between organisms according to the evolutionary distance between them (a measure of the ancestral development, hence evolutionary developments) and the results are presented as trees, as with family histories (see Box 6.1). Thus phylogenetics seeks to provide a classification that reflects the 'natural' relationship between organisms and has explanatory power for the development of this order.

---

■ **BOX 6.1 MOLECULAR CLOCKS**

Gene sequencing provides the molecular data for phylogenetic studies to compare the relationships between prokaryotes and eukaryotes. One particular concept that relates molecular data to evolutionary development is the idea of the molecular clock. The accumulation of mutations in the genes of certain marker molecules (notably ribosomal RNA) is assumed to accumulate at a fixed rate. If so, and this idea is not without its problems, then the number of mutations provides a molecular chronometer, an indication of the time elapsed. By comparing the number of mutations between different organisms it is then possible to estimate when the two organisms diverged (the 'evolutionary distance' between them). The greater the extent of mutation in the gene sequence places the organism as a more recent derivative. Conversely, the greater similarity in rRNA sequence between two organisms reflects a shared evolutionary history.

The rRNA molecule has been chosen because it is widely distributed across all organisms where it performs the same function and is also large enough to contain sufficient information for comparative analysis (see Box 6.2).

---

Figure 6.1 shows how the results of phylogenetic studies of molecules such as rRNA indicate that all life is encompassed within three *domains*:

**Bacteria, Archaea** and **Eukarya.**

This contrasts with the traditional view of the five *kingdoms* of life:

**Bacteria (or Monera), Plants, Animals, Fungi** and **Protists**

where only the first is prokaryotic, the rest being eukaryotes. Not surprisingly, such fundamental changes in the origins of life have led to some debate between biologists and microbiologists because the new three-domain system places bacteria as the ulti-

• **Figure 6.1** The phylogenetic tree of all life, based on comparison of rRNA sequences. The three Domains are shown in capitals. Note how animals, fungi and plants are all recent divergences and that Bacteria and Archaea are distinct

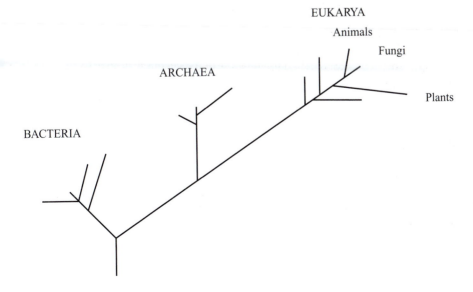

mate ancestor and all eukarya as late descendants. Most biologists will not take kindly to being told that they are a sub-discipline of microbiology! More pragmatic problems have arisen. Viruses have not been included since they do not code for rRNA. Nevertheless, to ignore such a large group of organisms is seriously misleading. Incorporation of viruses and related agents has been suggested at the level of a fourth Domain representing acellular infectious agents.

Phylogenetic classifications cut across conventional bacterial classifications. The rearrangements necessary to bring both systems of classification into line will result in tears as breaking with the old conventions will be difficult. So, far from dull regulations concerning name changes, taxonomy (systematics, if you prefer) has emerged as a catalyst for change with many intellectual challenges.

Conventionally, taxonomy comprises three topics:

- **classification**,
- **characterisation/identification** and
- **nomenclature**.

**Classification** is the theory and practice of ranking units (bacteria, viruses, fungi) into groups that can be ranked according to increasing inclusiveness (i.e. Species are grouped into a higher rank of Genus. Genera are included in a Family, etc.). Classification seeks to provide an ordered summary of the state of objects with predictive power. This means the properties of any one particular **species** or **genus** (etc.) will apply to your newly identified organism. Everyone expects a kilogram to contain 1000 grams, and in the same way one should be able to predict the properties of the new member of the group.

**Characterisation/identification** is the comparison of unknown units with previously recognised units/groups. True taxonomists consider identification to be a separate, distinct process that is not strictly taxonomy. In taxonomic terms, the process of obtaining information about organisms is called **characterisation**. The difference may appear slight but the characterisation process is carried out with a taxonomic objective, whereas identification is usually for a specific purpose in an applied setting such as the need to identify organisms recovered from clinical samples obtained from patients. The tests carried out in order to characterise an organism for taxonomic purposes will differ in scope and number quite considerably from those used for identification purposes. The characters tested are either **phenotypic** or **genotypic**. Phenotypic traits are observable physiological or biochemical properties that can be measured without any reference to its genetic basis. Hence, shape, number of flagella, ability to ferment sucrose, etc. are all phenotypic properties. Genotypic properties are molecular comparisons of nucleic acid such as gene sequences, base ratios, etc.

**Nomenclature** is the labelling of units into appropriate taxonomic positions as defined by classification.

This somewhat unfriendly series of definitions only formalises the actions that are carried out routinely in sorting out our books, clothes and weekly food stocks.

## ■ 6.2 CLASSIFICATION

Classification brings order to what otherwise would be a bewildering collection of objects. Now, some may argue that the constant revision of the classification of microbes does little to establish order but merely continues to destabilise; but it is the objective of classification to bring order and perspective. If such restructuring were to be avoided, one could just give a new number to each subsequent organism thereby simply 'cataloguing' organisms as we describe them. With such a system one could arrange them alphabetically or numerically.

As many different classification schemes have been described as there are characters. Cars can be classified according to size, colour, make, hub cap diameter, etc., but micro-organisms suffer in lacking many obvious morphological features. In addition, the embryological development of higher organisms can be used to help in their classification but again bacteria lack such clues. Nevertheless, ordering of microbes into basic groups can be carried out according to criteria such as biochemical properties or genome analysis. To help set a framework for comparison, **type strains** are held in reference collections of bacteria. These strains represent the typical organism that all species so named must closely match. All strains deemed to be sufficiently similar to this can be assigned to the same species and have the same name.

Classifications will always be artificial constructions because they are attempting to divide the spectrum of organisms that is actually a continuum. Fortunately, bacteria tend to fall into groups that can be distinguished from each other. The organisms themselves will evolve and change and this muddies the distinctions between them. This variability is in complete contrast to the ideas of Linnaeus who considered species to be distinct and unchanging, just like the stuffed animals in boxes in museums.

Classification orders organisms into species and then arranges species into increasingly larger taxonomic ranks such as genus, family, order, etc. From this process it is easy to view organisms inherently existing as distinct species. This idea is, however, mistaken. The concept of classifying organisms is a theoretical, abstract idea. Whilst a particular organism exists as a real phenomenon, classifying it as a particular species within a genus is an entirely conceptual event.

All biological classifications are comparative. The extent to which organisms differ determines the level at which they are grouped into taxa, such as families, genera or species. The species is the basic unit of division. What constitutes a species is another matter entirely and a source of endless discussion for taxonomists. Traditionally, it was possible to define a species by two ways:

- a **morphological species concept** (based solely on morphological characters), or
- a **biological species concept** (organisms able to breed (sexually) only amongst themselves. Higher organisms acquire their genetic material, in roughly equal proportion, from each parent. The genetic reassortment that occurs in meiosis will create slightly different genotypes for each individual. In addition, animals often are limited to geographic areas).

Both have obvious limitations when dealing with bacteria and viruses. Fungi are more amenable to such concepts and great emphasis has been placed on morphological structures in fungal taxonomy, but the complicated and varied sexual/asexual reproduction of different fungi make the biological species concept unworkable. Likewise, difficulties arise with bacteria because of genetic exchange. Even with parasexual events such as conjugation, genetic exchange can occur across genera as well as species. Hence, a third concept has been proposed: the **phylogenetic species concept**. As discussed above, the evolutionary history of micro-organisms has become one of the most important defining characteristics upon which the term 'species' has been built. The phylogenetic species concept views organisms as lineages and orders organisms according to the relatedness in protein and gene sequence patterns over evolutionary timescales. The number and type of differences in the sequences of certain marker proteins such as ribosomal RNA give an indication of how related the strains are, in evolutionary terms. The tracking of the rate of evolution in primary sequences means that the term 'phylogeny' is often interchanged with 'genealogy' because of the interest in the ancestors. Given the significance in variation in the primary sequences of the tracked proteins, it is necessary to make sure that the chosen proteins are common to all bacteria and perform the same function. 16S and 23S rRNA molecules fulfil these criteria for bacteria, 18S rRNA are used for fungi (see Box 6.2).

Whilst phylogeny seeks to uncover the universal tree of life, as sought by Darwin, bacteriologists are more often interested in less high-minded desires, driven instead in the search for order amongst the spectrum of properties exhibited by bacteria. These differences highlight how the concept of the species has dual function, depending on the user. For phylogeneticists, the species is the unit of evolution and therefore a variable entity, whereas for taxonomists, the species is a unit of classification that, to be of value to all biologists, needs to be as stable as possible.

For bacteriologists, a bacterial species is considered to be **a group of strains that share many common features but differ significantly from other strains**. The increasing analysis of microbial genomes has led to the concept of bacterial species as strains with high similarity in genomic relatedness (>70 per cent DNA sequence) as well as phenotypic characteristics. This reflects the idea that genes are exchanged in the natural world, hence species share a common gene pool. Ideally, the bacterial species should be based on the results of as many characters as possible and include phenotypic and genomic data. This inclusive approach is called **polyphasic**. A related concept is the **polythetic** species. Rather than choosing one particular property of an organism to be the defining property of the species, polythetic species have a set of characteristic properties that are shared amongst the species. The traditional view is to define a particular property that must be present in all members of that species, whereas polythetic classification permits inclusion of members that do not have one defining characteristic but a pool of shared characteristics (Figure 6.2). In this way individual organisms are not excluded on the basis of one result.

If the concept of species is woolly then the next taxa up is worse still. Species are collected into a genus (plural, genera). A genus is the collection of species that form a group sufficiently similar in themselves but sufficiently distinct from other (unrelated)

---

### ■ BOX 6.2 CHOOSING A MOLECULAR CLOCK

Genomic analysis of microbes has one clear advantage over traditional methods: the study of organisms that have not been grown in culture. Nucleic acid hybridisation will still occur in the organism and will be considerably faster in obtaining an identification. The choices of probe sequence has great importance and the target of choice is usually dominated by the ribosomal RNA (rRNA) sequences, either the genes coding for the rRNA (rDNA genes) or the ribosomal RNA itself. The bacterial ribosomes consist of the small 30S and large 50S subunits, which have 16S rRNA and 23S rRNA strands respectively. The rRNA is abundant in bacterial cytoplasm, are of appropriate base length and stretches are conserved within species (i.e. minimal mutational events can occur without serious detriment to the organism). Lateral transfer of these genes has not been a frequent event in evolutionary terms, thus they make a good target for estimating bacterial diversity and evolution.

• **Figure 6.2** Polythetic concept of taxonomy. Four bacteria are shown, each with their particular set of characters named A to E. If the species has been defined by the five properties A to E, then all four bacteria are members because, despite the fact that they do not possess all five properties, no one property defines the species. In this way individual strains that show variability are not excluded

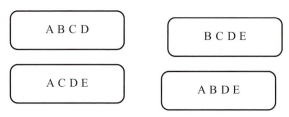

strains. The other ranks of taxa (Family, Order, Class) are broader, more inclusive levels of similarity. The order that different taxonomic ranks are arranged (in descending order towards species) is as follows:

Domain (Bacteria, Archaea, Eukarya)
Kingdom
Phylum/Division
Class
Order
Family
Genus
Species

Examples of three different micro-organisms are given in Table 6.1.

Microbiologists will talk mostly of species and genera. The higher ranks are usually only discussed by pure taxonomists who wish to make major rearrangements of the classification of organisms. With the increasing understanding of evolutionary history of microbes, discussions of this sort might become a more frequent event but at this stage we will stick mostly to genera and species. With fungi the taxon 'Class' is a useful tool, but somewhat frustratingly, viral classification has not strictly followed the binomial nomenclature, and genus and species names are not necessarily adhered to. Indeed, viruses are not represented in the highest taxa (Domain, Kingdom, Phylum or Class) but are only ranked from Order downwards.

**Table 6.1 Examples from the three micro-organisms representing bacteria, viruses and fungi**

| *Escherichia coli* | *Poliomyelitis virus* | *Cryptococcus neoformans* (sexual state: *Filobasidiella neoformans*) |
|---|---|---|
| Domain: Bacteria | | Domain: Eukarya |
| Kingdom: Bacteria | | Kingdom: Fungi |
| Phylum/Division: Proteobacteria | | Phylum/Division: Basidiomycota |
| Class: Zymobacteria | | Class: Basidiomycetes |
| Order: Enterobacteriales | Order: Nidovirales | Order: Tremellales |
| Family: Enterobacteriaceae | Family: Picornaviridae | Family: (not applicable) |
| Genus: *Escherichia* | Genus: Enterovirus | Genus: *Filobasidiella* |
| Species: *coli* | Species: *poliomyelitis virus* | Species: *neoformans* |

Classification systems in microbiology can be grouped into three types:

(i) **phenetic** classification: the natural grouping of organisms according to the maximum observable similarity of their characteristics (phenotype).
(ii) **phylogenetic** classification: the grouping of organisms according to the closeness of their evolutionary history.
(iii) **special purpose** classifications that are created for particular areas of microbiology such as food, medical or plant microbiologists.

Problems arise through conflicting interests of the different users which results in a lack of consensus of a single unifying classification system. Whilst a 'true' classification scheme is an arbitrary concept, the different systems are continually under revision through the continual fine tuning of the methods and results, thereby perpetuating a sense of uncertainty.

## ■ 6.3 CHARACTERISATION/IDENTIFICATION

The clumsy dual subtitle recognises that the identification of organisms from patients is not a true taxonomic process. Taxonomists will characterise isolates that are previously unrecognised in order to assign them to taxonomic groups (taxa) and not so as to help identify the cause of a patient's fever. Likewise, an organism can only be identified if it has previously been studied and given a taxonomic position. A semantic point perhaps, but worth consideration.

Two approaches can be adopted when identifying organisms, both illustrated in Figure 6.3. **Sequential identification** employs a series of different tests that determine the next course of action and are called **dichotomous steps**. The term 'dichotomous' should serve to indicate that, according to the result of one test, the identification can proceed one of two ways, only one of which is truly correct. A dichotomy that some might feel is too much like Russian roulette. The practice of proceeding on the basis of one test at a time has not been adopted in laboratories. The construction of suitable contrasting parameters is difficult and the step by step procedure prolongs the time taken to reach an identification as any one or more of the tests may require overnight incubation.

The **simultaneous identification** concept involves the use of a combination of several tests performed simultaneously. The results of all the tests are then checked against reference tables. By using a set of tests at once, the time involved is considerably reduced as is the emphasis given to each individual test by weighing up the total pattern of reactions. Simultaneous identification systems is more in keeping with a polythetic concept of a species than dichotomous key methods.

In practice, bacteriologists for example make a compromise between the two approaches in that a number of simple tests (e.g. Gram stain, motility, catalase and oxidase reaction) are carried out before setting up a more complete set of tests. This way the identification is optimised in terms of speed and cost.

For a list of tests that are used in the identification of bacteria see Barrow and Feltham (1993). The wide range of phenotypic tests measure a range of characters which include physiological (e.g. whether the organism is motile or not) or biochemical (e.g. the presence or absence of particular enzymes) characters. Table 6.2 lists the broad range of categories that can be used. Unfortunately, phylogenetic studies have shown that many of the traditional properties of bacteria that are used to create the current classifications (biochemical tests, gaseous requirements, Gram reaction, spore formation) are very poor markers of 'true' phylogenetic history.

• **Figure 6.3** (A) shows the procedure by which an organism can be identified using dichotomous keys. Depending on the result of the test, a particular series of tests are carried out until sufficient information is acquired to identify the organism. (B) shows a table where four organisms can be distinguished after testing six tests simultaneously. The letter V means that a variable proportion (between 20–80 per cent) of strains possess this property

(A) Sequential identification

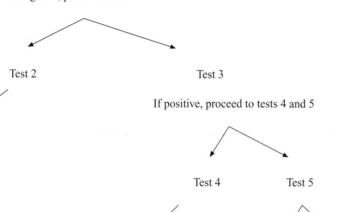

Test 1

If positive, proceed to test 2
If negative, proceed to test 3

Test 2                    Test 3

            If positive, proceed to tests 4 and 5

                    Test 4          Test 5

(B) Simultaneous identification

|            | Test 1 | Test 2 | Test 3 | Test 4 | Test 5 | Test 6 |
|------------|--------|--------|--------|--------|--------|--------|
| Organism 1 | +      | +      | +      | −      | −      | +      |
| Organism 2 | −      | +      | V      | +      | −      | +      |
| Organism 3 | −      | +      | −      | V      | −      | −      |
| Organism 4 | −      | −      | −      | −      | +      | +      |

For special purpose classification schemes, identification is usually the prime objective. The experienced microbiologist will determine which characters are of greatest discriminatory and predictive value, i.e., will give unequal weighting to different tests. The tests used to identify the organism will be very heavily weighted so as to optimise speed with accuracy. It cannot be overemphasised, however, that an accurate identification requires that the culture is pure.

## ■ 6.4 NOMENCLATURE

The name of someone imparts a certain amount of information simply through the agreed convention of carrying the father's surname down the male line. The names themselves can reveal something of the person's origins. Smith, Cheng, Gonzalez and Goldstein can suggest a certain country of origin. This is unlikely to be correct all the time, but it is a useful starting point. Similarly, the names can tell something of their function, for example, surnames derived from the person's occupation.

The naming of organisms is regulated by an international body which publishes the

**Table 6.2 Characteristics used to characterise bacteria**

| Category | Examples |
|---|---|
| Morphology of individual cells | shape, size, motility, sporulation |
| Cultural morphology | culture morphology: size, colour |
| Physiological properties | gaseous requirements, temperature and pH range for growth |
| Biochemical characters | utilisation of nutrients (especially carbohydrates), production of metabolic by-products (e.g. volatile fatty acids), presence of specific enzymes |
| Chemotaxonomic chemical analysis | Cell wall composition (peptidoglycan structure, teichoic acids, lipopolysaccharide structure) Cell membrane constituents (polar ester lipids, fatty acids, ether lipids, isoprenoid quinones, cytochromes) |
| Serological analysis | cell wall antigens |
| Genomic analysis | nucleic acid probe hybridisation, G:C ratios (see Box 6.1) |
| Inhibition profile | susceptibility to chemicals, antibiotics, phages |

accepted rules (International Code of Nomenclature of Bacteria, ICNB) and the approved new names. In this way it is hoped that duplication of names for the same organisms does not occur.

The **binomial** system of Linnaeus has been adopted by bacteriologists if not virologists, hence *Escherichia coli* is a bacterium within the genus *Escherichia* of which there are several species, *Escherichia coli* being the most familiar. The plural of species is often abbreviated to '*spp.*' as in *Salmonella spp.* Scientific names mostly concur with Latin grammar such that masculine, feminine and neuter words (names in this case) are predetermined, as shown in Table 6.3. When dealing with bacterial genera, the same rules apply. Hence we may have a single staphylococcus, clostridium or shigella but many staphylococci, clostridia and shigellae.

Taxonomists have made some effort to try to retain the older, more familiar names in an attempt to not disenfranchise all the end users like doctors and other health workers.

### ■ 6.5 NUMERICAL TAXONOMY

As more and more tests have been described for particular bacteria, the identification of a new bacterial species has involved characterisation across an increasingly broad battery of tests. Partly in response to the logistics of this sort of exercise, a more objective means of analysing and comparing phenotypic data was devised, known as

**Table 6.3 Latin-based grammatical construction of scientific terms**

| | Masculine | Feminine | Neuter |
|---|---|---|---|
| Singular | -us | -a | -um |
| e.g. | fungus | alga | bacterium |
| Plural | -i | -ae | -a |
| e.g. | fungi | algae | bacteria |

'numerical taxonomy'. Numerical taxonomy is often referred to as **phenetics** (and occasionally as 'Adansonian taxonomy' in recognition of the botanist who established the method). The principles of numerical taxonomy are:

- the taxonomic groups (taxa) are based on the greatest possible amount of information,
- the characters tested have equal weight,
- the similarity between organisms is directly related to the number of characteristics that are the same,
- the similarity of organisms is created independent of the evolutionary history or lineage of the organisms.

The basic strength of numerical taxonomy is the equal value (**weighting**) given to as many different tests as possible in order to determine the degree of similarity between organisms. The similarity coefficients obtained represent the proportion of characters that are similar between organisms. The overall similarity between organisms is next compared such that they are arranged into groups to separate dissimilar organisms from those that can be grouped together into clusters. In other words, the organisms are grouped according to their similarity between zero (no similarity) and 100 per cent similarity. Each cluster may constitute a distinct species (Figure 6.4). Numerical

• **Figure 6.4** Results of numerical taxonomic study of a collection of organisms expressed as a dendrogram. The '*x*-axis' is the percentage similarity between the strains tested. All the organisms with greater than 90 per cent similarity are collected together as numbers (right-hand column) that are equivalent to a species. For example, Group 1 would be *Staphylococcus epidermidis* and Group 2 would be *Staphylococcus aureus*. Where to define the levels of genus and species (as shown by the vertical dotted lines) will depend on the method used (e.g. DNA hybridisation). In this example, genera are defined by similarity of 80 per cent and species at 90 per cent

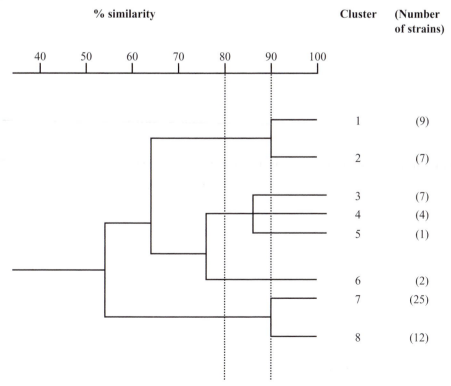

and H antigens after those in members of the Enterobacteriaciae) will give the particular strain a serotype. For example: *Esch. coli* O157; H15 will be the serotype. All isolates that are found to possess the O157 antigen form a serogroup. It follows then, that if only the O antigen is known, then the organism can only be typed to the level of serogroup because there will be more than one strain that has the same O antigen. Only when the H antigenic type is determined can the strain be referred to as a serotype.

### 6.6.1.3 Bacteriophage typing

Many bacteria have been typed by their pattern of lysis when challenged with a standardised set of lytic phages. A lawn of the organism on the surface of an agar plate is inoculated with a drop of the phage suspension. After incubation the organisms that have been lysed show a zone of no growth (a plaque). By comparing the obtained result with the strain under investigation with recognised patterns of lysis, a phage type can be obtained. Unfortunately, if there is no match with the recognised patterns then the method has limited value. Phage typing is normally carried out in national centres so as to keep a unified system. This puts certain restrictions on its usefulness and limits the number of organisms that have phage typing systems available. Examples of organisms typed by phage typing are *Salmonella typhi* and *Staphylococcus aureus*.

### ■ 6.6.2 NUCLEIC ACIDS AS TARGETS FOR MICROBIAL TYPING

Phenotypic tests have a number of limitations that reduce their suitability. The range of phages may be too small, the organisms may not have the phage receptors, the organisms have a restricted range of serotypes, etc. Unlike phenotypic methods, genotypic methods are usually guaranteed because all organisms have genes and therefore nucleic acid, even if it is RNA such as RNA-viruses.

Analysis of nucleic acid often separates varying lengths of DNA by gel electrophoresis to yield bands of different sizes. To create lengths of DNA of varying size, restriction enzymes are used. The different sites at which they cut the nucleic acid is the cause in variation of nucleic acid length. To add a greater degree of specificity, particular sequences of nucleic acid can be sought with the use of specific hybridisation probes.

The probes are applied after the bands have been transferred from the gel to a nitrocellulose membrane (the Southern blot) and visualised by coupling fluorescent or enzyme–substrate complexes to the probe. From this procedure, the technique called 'pulsed-field gel electrophoresis' (PGFE) has developed and is considered one of the best typing methods. Rather than look for specific sequences or genes, PGFE relies on a series of patterns formed by cutting the nucleic acid at rare (infrequent) restriction sites (hence 'unusual' restriction enzymes are employed). Because the sites are scarce, the bands obtained are usually large and these are best separated on gels by periodically changing the polarity of the current (hence 'pulsed-field') that carries the nucleic acid through the gel (Figure 6.5). The bands are visualised using ethidium bromide.

A variety of molecular typing systems have been described, many of them employing polymerase chain amplification of the target sequence before separation by electrophoresis. The differences in methods centre on the choice of DNA sequence that is chosen for amplification.

• **Figure 6.5** Pulsed-field gel electrophoresis. The bands obtained from four strains of the same species, illustrating that strains 1 and 2 are identical and thus considered to be the same strain, whereas strains 3 and 4 are distinct

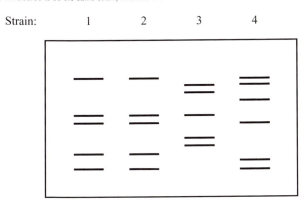

Strain: 1 2 3 4

## ■ SUMMARY

Taxonomy comprises the principles, methods and rules that govern classification, characterisation and nomenclature. The identification of bacteria is often considered a distinct exercise from characterisation because the methods used are not sufficiently rigorous. Bacterial classification has relied on the grouping of bacteria according to the patterns of characters they exhibit (phenotypic characters) that have been analysed using numerical taxonomy. With increasing emphasis on genetic methods by sequencing genes a mixed phenotypic and genomic (polyphasic) approach has become possible. However, methods of obtaining the evolutionary history of bacteria (phylogenetics) have posed problems by failing to match the old classification systems. Phylogenetics is thought to represent the 'natural' pattern of relationships between bacteria by comparing the gene sequences of conserved (slowly evolving) proteins or nucleic acid (typically rRNA). Such studies have revolutionised the old classifications of life and placed bacteria at the root of the evolutionary tree. Phylogenetics has meant that bacterial taxonomy has moved from descriptive to explanatory and, as a result, the concept of 'species' now serves two functions: a unit of evolution and a unit of classification.

Typing a set of strains of a particular species requires a pattern of results that can be compared between the different strains. Whereas identification uses tests that ideally yield a yes or no answer, typing demands a certain degree of complexity in the results if adequate discrimination is to be obtained. Typing relies on the test yielding a range of patterns within a random collection of strains. Only then is one able to impart significance to a repeated or common pattern within a collection of strains.

## RECOMMENDED READING

Balows, A., Truper, H.G., Dworkin, M., Harder, W. and Schleifer, K.H. (1991) *The Prokaryotes. A Handbook on the Biology of Bacteria*, 2nd edition, Springer, Berlin.

Barrow, G.I. and Feltham, R.K.A. (1993) *Cowan and Steel's Manual for the Identification of Medical Bacteria*, 3rd edition, Cambridge University Press, Cambridge, UK.

Cowan, S.T. (1971) Sense and nonsense in bacterial taxonomy. *J. Gen. Microbiol*. 67, 1–8.

Garrity, G. (2001) *Bergey's Manual of Systematic Bacteriology* (No.1), 2nd edition, Springer-Verlag New York Inc., New York, USA. (See the updated introductory chapters.)

Holt, J.G., Kreig, N.R., Sneath, P.H.A., Staley, J.T. and Williams, S.T. (1994) *Bergey's Manual of Systematic Bacteriology*, 1st edition, Williams and Wilkins, Baltimore, USA. (See the introductory chapters on bacterial classification.)

by more than one mode of transmission. Measles virus is likely to be spread by airborne and direct contact. The number of times people touch their faces with their hands is enough to almost guarantee that virus in the nose or saliva will be transferred to the hands. You will notice other flaws in the scheme. The table distinguishes between sexually transmitted infections and those obtained by direct contact. It could be argued that they are the same mode of transmission. The example of dermatophytes as directly transferred infections also needs qualifying. In the latter cases the fungi are often transmitted through shared, inanimate objects like towels or combs. This is not direct transmission in the strictest sense. Perhaps 'indirect transmission' is a more appropriate term.

Two important terms are **vertical transmission** and **horizontal transmission**. Rather than referring to positions of the host when they came into contact with the infectious agent (!) they are used to distinguish between infections transferred from mother to child (**vertical**) and those acquired from other sources (which in this context means predominantly people) and are called **horizontal**. Two of the most important examples of vertical transmission are rubella virus and *Treponema pallidum* (syphilis). Such infections of the foetus can have serious developmental consequences. Congenital rubella, for example, may lead to microcephaly (small head), impaired vision and impaired intellectual development.

The significance of vertical or horizontal transmission for the microbe lies in the number of people that can be infected. With vertical transmission only one new host is infected and that is the child (or children). With horizontal transmission there is the opportunity to infect significantly greater numbers of people. It follows then that if one person is able to infect five others, these five will then infect 25 and so on.

The exact point at which infections of the newborn baby stop being vertical transmission and become horizontal is debatable. Certainly vertical transmission can be subdivided into:

- germline transmission (spread of virus integrated within the genome),
- prenatal transmission (infection of the foetus in the uterus),
- perinatal transmission (infection during birth),
- postnatal transmission (infection following birth).

Strictly, only infections of the foetus *in utero* are vertically transmitted. Perinatal infections (those acquired by the baby in the vaginal tract during the process of delivery) and postnatal infections (e.g. acquired via breast milk) are horizontally transmitted.

**Vector-borne infections** are those that are transferred by a vehicle/carrier, which are insects. The term vector is (most usefully) restricted to living animals such as lice, fleas or mosquitoes. Combs and hairbrushes are thus not usually called vectors, but fomites. The transfer of dermatophytes via hairbrushes is best described as 'direct transfer'. Vectors serve to transmit organisms from the bloodstream of infected patients to new susceptibles. The organisms (notably **arboviruses**: *ar*thropod-*bo*rne viruses and protozoan parasites such as malaria) are not designed to withstand any time outside of a host. The transfer of the organism is dependent on the distances the vector can travel in its lifetime. A wide range of insects transmit infectious microbes, as shown in Table 7.6.

**Table 7.6 Insects that transmit infectious micro-organisms**

| Insect | Disease | Microbe |
|--------|---------|---------|
| Bugs | Chagas' disease | *Trypanosoma sp.* (protozoa) |
| Fleas | Plague | *Yersinia pestis* (bacterium) |
| Lice | Typhus fever | *Rickettsia sp.* (bacterium) |
| Ticks | Lyme disease | *Borrelia sp.* (bacterium) |
| Flies | Sand fly fever | sand fly virus |
| Mosquito | Yellow fever | yellow fever virus |

## ■ 7.7 ZOONOSES

Zoonoses are infections that can be transmitted between vertebrate animals and humans (singular: **zoonosis**). The natural host is the animal. Often a human infection represents a dead-end in terms of perpetuating the spread of the organism because we are unlikely to pass the organism to another animal.

Having mentioned malaria as an arthropod-borne infection, one should note that malaria in man is *not* a zoonosis. There are numerous (hundreds) of animals that are infected with different *Plasmodium spp.* but the plasmodia that commonly cause disease in humans appear to have evolved so that humans are the 'natural' hosts, thereby disqualifying themselves as zoonoses.

---

### SOURCES OF ZOONOSES

Zoonoses have arisen mostly through the need of humans to collect food. Humans have been able to domesticate animals both as livestock (cows, pigs, etc.) and as helpers (sheepdogs) as well as deforesting areas for crops. These activities bring man into close contact with the micro-organisms that infect the animals. Tuberculosis, anthrax and brucellosis are caught directly from farmed animals. Others arise from those animals we encounter whilst disturbing them in their natural habitats (yellow fever, Lyme disease, Ebola virus). Not that zoonoses are only caught by venturing into dark jungles! Pets are a common source of various microbes that can cause human disease, especially fungal dermatophyte infections. More important in terms of national public health is food poisoning. Increasing in incidence in the Western world, *Salmonella enteritidis* and *Campylobacter jejuni* are the leading causes of food poisoning in developed countries. Mostly as a result of battery farming, both organisms are frequently isolated from uncooked chickens but inadequate precautions in food hygiene fails to reduce their numbers to below the infectious dose necessary to induce food poisoning. Despite being a preventable infection, food poisoning is a national public health concern and burden in lost manpower to industry.

---

The transmission from the animal to humans (and it is only this direction that concerns us here!) can be separated further, depending on the natural lifecycle of the micro-organism in question.

1. Direct transmission (direct zoonosis): transmission occurs via direct contact from the original animal (e.g. rabies transferred through the saliva of the dog). The microbe does not undergo any developmental change between the two hosts (rabies, brucellosis).

organisms found in these sites is considerably higher than on the skin, probably due to the degree of available water. The mucous lining covering the epithelia of these body cavities and tubes promotes a deep layer of organisms that arrange themselves according to the extent to which they are aerotolerant. At the surface of the epithelial cell, the $E_h$ will be highly negative (that is, very reduced, very anaerobic). The organisms will layer themselves so as to mirror the oxygen tension. The most oxygen sensitive will be closest to the tissues, and the more aerotolerant organisms will layer themselves in the outer, more oxygen-rich zones closer to the outer surface.

Hopefully the discussions of bacterial growth in Chapter 2 are beginning to ring some bells. The different sites of the host that are colonised with the normal flora can be seen to provide the requirements for growth. What provisions, then, does the body take to limit itself from becoming overwhelmed by proliferating microbes? Those people with poor dental hygiene amply provide an example of the balance between growth and growth limitation being lost. Several mechanisms are used by humans to control the numbers of organisms carried. The body loses squamous epithelia from the skin along with the associated bacteria. Coupled to regular shedding, the skin will dry out, thereby reducing favourable growth conditions. The skin also has several antibacterial agents. Lysozyme is secreted into sweat and the acids produced by certain skin bacteria (propionibacteria) are inhibitory to certain organisms, not least through dropping the pH. Rapid utilisation of the available nutrients by the flora will be a major limiting factor for bacterial growth in all sites. In the intestine, this is likely only to be a temporary starvation. To get to the intestine, bacteria pass through the stomach but stomach acid will help kill most bacteria. The numbers of bacteria in the proximal small intestine are low but increase the further you go down the intestine. Highest numbers are found in the colon. Tolerance to bile also helps. The differences between Gram positive and Gram negative bacteria in resistance to membrane destabilising agents such as bile salts was discussed in Chapter 1. Within the small intestine the epithelia are continually replaced as the tips of the villi and adherent bacteria are released into the lumen, to be carried out with the faeces.

Factors that limit bacterial proliferation are not all host derived. The competition for nutrients and space by other bacteria are also recognised. Certain bacteria produce **bacteriocins** when they are in late exponential/stationary phase of growth. These secreted proteins act with varying specificity to inhibit or even kill other bacteria. The elimination of competing species is clearly an advantage when the food returns. The space left by the killed organisms is also then available for occupation.

The normal flora does not appear to be simply parasitic or even cohabiting as commensals because there are a number of functions and benefits that humans derive from the normal flora. Again, the GI tract provides several examples of benefits to humans, with the essential digestive functions of the flora in the rumen of herbivores as a famous example.

- The synthesis of vitamin K and vitamin B12 by the intestinal flora. Germ-free animals (**gnotobiotic** animals) suffer from vitamin K deficiency until the intestinal flora is allowed to establish.
- The **short chain volatile fatty acids** (VFAs) produced by anaerobes in the colon are utilised by the epithelial cells lining the colon (colonocytes) as their primary metabolic fuel (over glucose). This makes sense in light of the absence of glucose that far along the intestine. VFAs as nutrient supplies to the colon is highlighted when the colonic bacteria are reduced in number in patients with a colostomy that

develop a severe colitis in the defunctioning length of colon (from stoma to anus). When the terminal length of colon is flushed with saline in an attempt to keep it clean, the starvation of the colonocytes results in colitis (called **diversion colitis**) which is reversed by passing a solution of the VFAs into the lumen. The substance that imparts the distinctive odour of faeces is the very stuff that is used to feed the intestinal epithelia.

- A third example of benefit is the finding that gnotobiotic animals have poorly developed lymphoid tissue compared with the normal animals. The continued exposure to bacterial products (cell wall especially) acts as a continual low-level stimulus for antibody production in the gut associated lymph tissue.

## ■ 7.8.2 PROTECTION AGAINST PATHOGENS

The normal flora offers some protection against invading pathogens. The pattern resembles the problems of getting seeds to root in an established field of grain. The space is limited, the water and soil nutrients are all being taken by established root systems. This protection, when applied to the intestine, is called **colonisation resistance**, although the principle applies to any situation on the human body where the normal flora limits the ability of an invading organism to gain a foothold. The efficacy of this protection is highlighted by the disease **pseudomembranous colitis** (PMC). Patients with PMC develop severe diarrhoea and have a characteristic membrane over the surface of the colon visualised by sigmoidoscopy. PMC is caused by the multiplication and colonisation in the colon by the anaerobic Gram positive rod *Clostridium difficile*. When the normal intestinal flora is disrupted by treatment with poorly absorbed antibiotics, patients can become readily colonised by *Clostridium difficile* from sites vacated by the depleted normal flora. *Clostridium difficile* can produce at least two protein toxins that contribute to the diarrhoea that develops. The organism can be isolated from healthy people at low frequency, implying that the organism either exists in very low numbers in the gut and only multiplies when given the opportunity following antibiotic treatment, or is ingested at the appropriate time (in hospital, for example).

Colonisation resistance applies to all sites where the normal flora is present. Within the vaginal tract, glycogen is present in relatively high concentrations due to the resident lactobacilli that produce lactic acid as the principle by-product of fermentation of glycogen. This production of acid lowers the pH to acid levels as low as pH 5, an environment that is antagonistic to many bacteria.

Harbouring a normal flora is not without risks. Under normal circumstances the normal flora will most likely not present a problem. As soon as new and unusual circumstances arise, however, some microbes will seek to exploit a new potential site for growth and colonisation. Anyone who has cut themselves will recognise that often the cut becomes infected. Most likely it will be a *Staphylococcus aureus* infection, the organism originating from the nose. Urinary tract infections by *Escherichia coli* will arise from the organisms that have passed through the gastrointestinal tract. Such organisms are thus termed 'opportunists'; they require a breakdown in the normal protective mechanism (e.g. cut in the skin). It is also possible to culture organisms that cause serious infections from a person who is asymptomatic. We can conclude thus far by saying that the relationship between host and the normal flora is in equilibrium until a perturbation of normal flora or host tissue occurs. Then the balance shifts in favour of the bacteria, rapidly multiplying in the new environment, at least in the short term.

Knowledge of the dominant types of bacteria from the different sites of the body is essential when looking for pathogens in clinical samples. The use of appropriate selective culture media can be employed to help isolate any pathogens in a sample that may be dominated by normal flora (e.g. looking for salmonellae in faeces). The diagnostic microbiologist needs to learn how to recognise and distinguish the normal flora and the potential pathogen, not least because it is not always possible to obtain a specimen from the patient without sampling the normal flora as well as the diseased tissue. A patient with a chest infection needs to produce a sample by coughing. The difficulty is trying not to contaminate the sputum (from the lungs) with saliva and organisms from the normal flora of the mouth. These procedures are possible because the organism you are looking for is known and identifiable. Dental bacteriologists have not had it so easy. It is likely that tooth and gum decay is caused by the collective action of members of the normal mouth/gum flora. The problem arises because of the vast excess of glucose and other refined sugars that the modern diet provides. The marked acidity resulting from the fermentation of the sugar erodes the tooth enamel and enables a proliferation of anaerobic (hence, foul smelling) bacteria. It has been not been possible to convincingly identify a single culprit responsible for dental caries.

You will have noticed that the normal flora has been contained to bacteria and yeasts that live on the external faces of the body (where the respiratory tract, genito-urinary tract, etc. are considered external faces). Internal organs are considered sterile. It is tacitly accepted that the presence of viruses is abnormal (i.e. an infection). The sequencing of the human genome has unearthed numerous stretches/fragments of past virus infections. Whether viral genome in human tissues will ever be considered normal flora is up for debate.

One important area of concern to hospital microbiologists is the problems of infections transmitted in hospitals. The topic brings together both the transmission of microbes and the state of the health in the host. Does illness cause a shift in the balance between the host and the microbes they encounter?

## ■ 7.9 NOSOCOMIAL INFECTIONS

Florence Nightingale said that going into hospital should at least do the patient no harm. Today, people often acquire an infection during their stay in hospital and these are called **nosocomial** infections. The reasons are not hard to find. Patients will likely have one or more of their natural protective mechanisms breached. This can mean inserting an intravenous line or undergoing surgery, thereby breaching the intact skin barrier. Alternatively, if not in addition, they might have a urinary catheter inserted, thereby compromising the normal flushing by passing urine via the urethra. The opportunities for opportunistic microbes to establish an infection are provided. To further increase the risk, the patient will be visited by numerous staff and they themselves get the opportunity to touch and collect bacteria from all the other patients. Despite rigorous hand washing it is not always possible to remove all the bacteria. Nosocomial infections are opportunistic infections, considered separately from those infections acquired in the community. The distinction is usually made because the compromised condition of the patient leads them to be become infected with micro-organisms that, in healthy people, present little problem/risk. Not that hospitals are protected from ordinary infectious diseases. The infectious diseases that infect healthy people outside of hospitals can easily be brought into hospitals via the staff.

Infections that are acquired from external sources, i.e. are not normally resident within the host, are termed **exogenous** infections. Those infections that originate from microbes present in the host normally are called **endogenous**. The terms 'exogenous' and 'endogenous' infection have particular value when investigating nosocomial infections because many infections arise from the patient's own flora. If the infecting agent can be shown to be an exogenous infection it points to a problem of cross-infection. The term **iatrogenic** infection applies to those infections that arise as a direct consequence of a medical or surgical procedure (e.g. infection of a new hip joint, infections following insertion of a urinary catheter or intravenous line). Iatrogenic infections, for example an infected intravenous line, can be either endogenous or exogenous in origin. The implications for control are obvious, the cost is another factor altogether.

The sources of nosocomial infections are other people (staff or patients) or the patient's environment. This will include such diverse sources as organisms growing in ventilator tubes, or inadequately disinfected endoscopes, through to organisms in the dust, air or food. In addition to the compromised defences of the patient, the bacteria that cause nosocomial infections are increasingly resistant to the action of antibiotics. Often, the organism is able to tolerate a range of different antibiotics (**multiple antibiotic resistance**), posing greater problems in the treatment of the infection.

It is interesting that bacteria are the most common causes of nosocomial infections. However, more recent trends show an increasing incidence in fungal infections. It is suggested that the advances in medicine are paralleled by an increase in the number of infections by organisms of low virulence. The people who previously would have died, who now are being kept alive, are liable to acquire infections with organisms correspondingly 'weak'. Whilst viral and fungal infections do occur, bacteria appear to be the dominant problem. Why?

The reasons may be the source of nosocomial infections: humans and contaminated objects. Humans carry vast numbers of microbes as normal flora, and bacteria are able to quickly grow to large numbers in a wide variety of sites. Providing moisture and nutrients are available, bacterial growth will occur. Being microscopic, they will not necessarily be dealt with. Fungi may be able to grow on a wider range of substrates with less moisture ($a_w$) but they form visible colonies of mycelia. Once noticed, they will be eliminated. As seen in Chapter 3, viruses will have an obligatory requirement for living hosts if they are to multiply. Viral infections might induce illness and this may limit their spread, as suitable precautions can be taken to prevent further infections.

The most frequent types of infection in hospitals occur in the urinary tract, bloodstream, the chest and in surgical wounds. The risk of developing a nosocomial infection is related, mostly, to the severity of the underlying disease. Patients who have impaired defences, whether they are inherited diseases of impaired immunity or secondary to other diseases or the treatment are all called 'immunocompromised hosts'. The efficiency of treating the infection corresponds to the resolution of the original pathology. In addition to the primary disease predisposing a patient to nosocomial infections, the treatment itself can often have an effect. Drug treatment to immunosuppress patients following transplantation or treatment of cancer with cytotoxic drugs will impair the host immune response. Antibiotic treatment itself will disturb the normal flora (and therefore the colonisation resistance) and can select for bacteria that are resistant to that antibiotic.

The interactions between populations defined at the start of the chapter tends to imply that they are stable in nature. More correctly, these states will continually be challenged by new circumstances which result in changes in the interactions. The dynamic nature of population interactions is highlighted by nosocomial infections where the condition of the host, the patient, is temporarily (we hope) altered. The medical and surgical improvements have created a greater variety of situations for microbial exploitation with tremendous financial costs. The challenge for medicine is to attempt to treat disease with as minimal disruption to the patient's physiology and normal host flora as possible.

From such pragmatic issues we conclude this chapter by revisiting the biology of the micro-organism and the reasons for parasitic infections in humans. What are the costs and benefits in choosing to infect humans?

## ■ 7.10 WHY ARE MICROBIAL PATHOGENS DAMAGING TO THE HOST?

For a long time it was thought that infections are damaging to the host because of insufficient time for adaptation by the microbe to the new host. In evolutionary time, the association between the microbe and the host has been too short to allow the organism to gradually reduce its virulence. The underlying rationale being that the microbe does not want to damage its host; in other words, all organisms are striving for a state of commensalism. The problem with this argument is that evolutionary pressures do not worry about the host; the primary importance is the propagation of the self, whatever the cost to the infected party. Current thinking favours the trade off between reproductive fitness of the organism and the survival of the host. The organism may damage the host in the quest for obtaining nutrients, but the long-term yields will be curtailed if the host dies. The trade off means that a balance will be established between being too damaging or not damaging enough (Figure 7.2) rather than just becoming harmless. There are a number of proposed theories that offer new views on the driving forces for virulence in microbes (and parasites in general). Note that 'virulence' here simply means damage to the host.

## ■ 7.10.1 VIRULENCE IS A RESULT OF THE ORGANISM INFECTING THE WRONG (UNINTENDED) HOST

The devastating effects of Ebola virus outbreaks are considered an example of such a hypothesis. Ebola virus is thought to be a zoonosis and the accidental infection of humans is a result of the virus straying from its natural host (as yet unknown). The virulence observed in the accidental host is considered irrelevant because the infection is not part of the normal path of transmission, e.g. *Cryptococcus neoformans* meningitis. You should note that such an infection yields almost no value to the parasite in terms of enhanced reproduction or transmission. By using Ebola virus infections, the hypothesis artificially suggests that all infections in accidental hosts will have high morbidity and/or mortality. This impression results from the high impact of 'virulent' infections such as Ebola virus. The reality may be that most infections in unnatural hosts fail to

• **Figure 7.2** The trade off between virulence and transmission for a micro-organism that can kill its host. The optimal level of virulence will be balanced at the apex of the curve to maximise both mortality and transmission

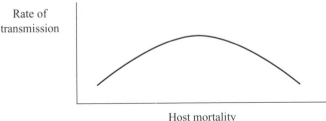

establish themselves and those that do are asymptomatic. We are left with wondering why strict human pathogens are virulent in humans, e.g. measles or typhoid. The hypothesis does not readily explain their virulence.

### ■ 7.10.2 BETTER TO BE VIRULENT THAN HARMLESS

Evolutionary pressures will favour virulence in a parasite if it results in increased reproductive success, albeit at the expense of the host. So, if any damage to the host is going to help the reproduction or transmission of the parasite (e.g. causes diarrhoea or coughing) then virulence is better than being harmless.

### ■ 7.10.3 VIRULENCE IS A TRADE OFF AGAINST OTHER CHARACTERISTICS

In this scenario, the organism causes harm to the host 'against its own wishes' because it is a necessary evil if the organism is to survive in the host. The virulence trait is offset by the advantage gained in the host. In the need to acquire nutrients, for example, the organisms need to lyse host cells to release the cell contents. This provides nutrition but causes tissue damage which, in turn, induces a host immune response.

### ■ 7.10.4 VIRULENCE IS DETERMINED BY THE ARMS RACE BETWEEN HOST AND PARASITE

This hypothesis arises from a view that both the host and the parasite are evolving in response to each other. If the parasite did not adapt then the virulent strains would be eliminated by the host and harmless strains would dominate. Virulence will decrease but the parasite might lose fitness, hence the parasite is proposed to counter the efforts of the host in eliminating it by developing new virulence traits to maintain its own fitness.

Much debate has arisen concerning the evolution of virulence. At this stage the evidence supports the idea that the interplay between parasites and their hosts is going to vary in each case. The interplay between damage and host response will result in a trade off between harmless and damaging depending on what is optimal for the parasite. If damage results in greater evolutionary fitness, greater reproductive success then being nasty is best. The study of the introduction of myxoma virus (causing myxomatosis in rabbits) into Australia has provided a useful case study. In 1950 the virus was introduced to control the burgeoning rabbit population. Over the next ten years, rather than become harmless, the virus evolved to an intermediate degree of virulence, presumably sufficient to maintain itself within the rabbit population. The outcome of parasite–host interactions will vary according to the circumstances, but evolutionary pressures mean that the parasite may adopt a harmless, virulent or intermediate strategy depending on the reproductive fitness.

## ■ SUMMARY

Bacteria can be seen as either generalists or specialists. Generalists will occupy a broad niche, where niche refers to the requirements of an organism rather than physical location. Specialists will be more exacting and limited in their capabilities and habitats. It is likely that the extent to which micro-organisms adopt mutualistic or parasitic modes of existence will depend on whether they are generalists or specialists. The communities of microbes that are resident on or in the human host (the normal flora) provide a useful casebook for studying the balance in human host–microbial parasite equilibrium. Benefits (e.g. colonisation resistance) and costs (endogenous infections) are implicit in maintaining a normal flora. The relatively few primary pathogens that infect humans are usually exogenously acquired and only transient parasitic episodes. Clinical disease often only develops in a fraction of the total number of people infected. The outcome will depend on the virulence of the organism and its infectious dose along with the condition of the host. The modes of transmission are varied, as are the sources and reservoirs of infection, but the majority of encounters will be via mucous membranes. Where it was previously considered that disease results from the organism having inadequate time to adapt to a lower level of virulence, recent evolutionary theory proposes micro-organisms adopt a level of virulence that is optimal for reproductive success and this includes efficient transmission between hosts.

## RECOMMENDED READING

Atlas, R. and Bartha, R. (1998) *Microbial Ecology: Fundamentals and Applications*, 4th edition, Benjamin Cummings, California, USA.

Balows, A., Truper, H.G., Dworkin, M., Harder, H. and Schleifer, K.-H. (1991) *The Prokaryotes. A Handbook on the Biology of Bacteria*, 2nd edition, Springer, Berlin.

Berg, R.D. (1996) The indigenous gastrointestinal microflora. *Trends Microbiol.* 4, 430–5.

Ebert, D. and Herre, E.A. (1996) The evolution of parasitic disease. *Parasitology Today* 12, 96–101.

Mason, T.G. and Richardson, G. (1981) *Escherichia coli* and the human gut; some ecological considerations. *J. Appl. Bacteriol.* 1, 1–16.

Mims, C., Playfair, J., Roitt, I., Wakelin, D. and Williams, R. (1998) *Medical Microbiology*, 2nd edition, Mosby International Ltd, London, UK.

Moran, N.A. and Wernegreen, J.J. (2000) Lifestyle evolution in symbiotic bacteria: insights from genomics. *Trends Ecol. Evol.* 15, 321–6.

Vollaard, E.J. and Clasener, H.A. (1994) Colonization resistance. *Antimicrob. Agents Chemother.* 38, 409–14.

## REVIEW QUESTIONS

*Question 7.1*    What is 'colonisation resistance'?

*Question 7.2*    To what extent does the normal flora of humans provide examples of commensalism, symbiosis, synergism, competition and parasitism?

*Question 7.3*    List the benefits and costs of humans housing a normal flora.

*Question 7.4*    List examples of microbes that are transmitted by 1) water, 2) soil, 3) food, 4) air(borne) and 5) blood.

*Question 7.5*    Distinguish between a source, a reservoir and a vehicle.

Fortunately, only a small fraction of the number of bacteria recognised are able to cause disease in humans and animals. What is it about this select group of organisms that confers the ability to cause disease in humans? This chapter will examine some of the properties of these bacteria.

Before discussing how pathogens behave, let us examine the criteria by which organisms are classified as pathogens or not. A patient has a small abscess on her hand. The lesion is red and inflamed and pus is draining from it. You culture the pus and you isolate an organism. How do know whether it is responsible for the inflammation? Did the organism cause the lesion or simply arrive and start multiplying after the lesion appeared? This sort of problem is conventionally addressed by reference to **Koch's postulates**.

## ■ 8.1 KOCH'S POSTULATES

In the last decade of the 1800s, the bacteriologist Robert Koch demonstrated that *Mycobacterium tuberculosis* was the aetiological agent of tuberculosis. Having grown the organism from human cases of tuberculosis on his newly created culture media, he succeeded in reproducing a tuberculosis-like disease after inoculating the organism into rabbits. Although never directly stated by Koch, a number of necessary steps have been outlined which are to be followed when one wishes to see whether a particular bacterium is the cause of a particular disease. They are called Koch's postulates and can be summarised as follows:

* the organism should be found in all cases of the disease, and in the affected lesions,
* the organism should be grown from the lesion in pure culture,
* inoculation of the isolated organism should reproduce the disease in a susceptible animal.

Koch's postulates were later amended to the following:

* the microbe must be present in every case of the disease,
* the microbe must be isolated from the diseased host and grown in pure culture,

- the disease must be reproduced when a pure culture is introduced into a susceptible host,
- the microbe must be recovered from an experimentally-infected host.

The logic cannot be faulted and in the late nineteenth century Koch was setting admirable standards with both the theoretical principles and practical methods for investigating infectious diseases. Today, however, they need qualification. Some of the postulates cannot be fulfilled because:

- we cannot culture certain organisms in culture media. To the embarrassment of medical microbiologists everywhere, the causative agents of syphilis (*Treponema pallidum*) and leprosy (*Mycobacterium leprae*) cannot be grown outside animals, i.e. on culture media *in vitro*.
- certain diseases are caused by combinations of bacteria. Dental caries is widely believed to a polymicrobial condition.
- many specific infectious diseases of humans have no appropriate or equivalent animal counterpart, hence the human disease cannot be reproduced. Examples include the sexually-transmitted diseases syphilis and gonorrhoea, for which no suitable animal models exist that mimic the human infection.

Furthermore, there are other issues that need comment.

### ■ 8.1.1 SITE OF ENTRY

The site and method by which an animal is infected with the test organism is not accounted for. Intraperitoneal injection is the standard site for transmitting *Mycobacterium tuberculosis* in guinea pigs. Whilst this method is very efficient in the transfer of the organism in these animals it bears little resemblance to the respiratory route of transmission in man. Likewise, mice are unaffected by inhalation of aerosols of encapsulate *Streptococcus pneumoniae* but very quickly succumb to intraperitoneal injection.

### ■ 8.1.2 INFECTIONS DUE TO SECONDARY PATHOGENS

Infections by organisms that are are unable to cause clinical disease in healthy people with no predisposing factors (immunosuppression, catheters and intravenous lines, etc.) were not studied by Koch. Koch's postulates were conceived whilst identifying what we now call 'primary pathogens' and the problems of identifying opportunist infections were not appreciated at the time. Since then we now appreciate that certain organisms are able to exist as commensals but at certain times proliferate when certain opportunities arise. The results obtained from injecting organisms into healthy animals give little clear indication as to whether the organism is acting as a primary or secondary pathogen. *Staphylococcus aureus* can readily be isolated from the noses of healthy people but has the potential to cause a huge variety of specific infections. Thus, simply isolating the organism from patients is not enough to decide whether it is responsible for any particular disease without considering the context.

### ■ 8.1.3 RISE IN ANTIBODY TITRE

With appropriate methods, one can measure quantitative increases in levels of specific antibody (increasing titre) against the pathogen, strong evidence for infection by that organism. The increase in antibody titre is important because the qualitative presence of antibodies alone does not distinguish between previous or current infection.

## ■ 8.1.4 CARRIER STATE

It has become clear that certain organisms are able to induce a carrier state. *Salmonella typhi* causes typhoid fever but in a very small proportion of cases persist in the gall bladder of people for years. They need not have developed the full clinical illness yet still excrete the organism in the faeces. The existence of asymptomatic carriers does not argue against *Salmonella typhi* causing typhoid but does complicate the interpretation that it is a primary pathogen.

## ■ 8.2 HOW DO YOU DEFINE THE ABILITY TO CAUSE DISEASE?

### 8.2.1 BACTERIAL PATHOGENICITY

The ability of an organism to cause disease in a host is referred to as the **pathogenicity** of an organism. The process by which the disease is caused is termed the 'pathogenesis' (composed from the phrase 'genesis of pathology'). The disease will be the result of the destructive effects of the bacterium itself *and* the effects of the host as it attempts to deal with the invader. Pathogenicity is therefore the sum of bacterial and host effects and not the sole action of the organism. Table 8.1 helps illustrate the concept of pathogenicity. Using the same inoculum size for both animals, the results show that Organism X is pathogenic for mice but not rats. To examine the role of the host in greater detail we will look at two organisms in the same host: humans. In Table 8.2 we compare the outcomes of exposing the same animal with two different bacteria. *Staph. aureus* is shown to be the more pathogenic of the two organisms tested because it was able to establish itself on the abraded skin and cause an infection, whereas the *Staph. epidermidis* was unable to do so under the same conditions. For *Staph. epidermidis* to establish itself, the presence of a foreign body was required (the IV line catheter). This illustrates the importance of the condition of the host. The condition of the skin had a large influence on the outcome following exposure to two different organisms but the invading organisms differed in their ability to exploit the abraded skin. Such an experiment tells us nothing about the mechanisms that account for the virulence of the organisms. The properties of the organism that were responsible for the difference between the two organisms are called **virulence** factors and further experiments need to be devised to test for virulence factors.

### Table 8.1 Effect of route on bacterial virulence

| Organism | Animal | Route | Outcome |
| --- | --- | --- | --- |
| Organism X | rat | inhalation | no effect |
| Organism X | mouse | inhalation | pneumonia |

### Table 8.2 Outcomes of inoculation onto skin

| Organism | Conditions | Outcome |
| --- | --- | --- |
| Staph. aureus | healthy skin | no infection |
| Staph. aureus | abraded skin | infection |
| Staph. epidermidis | healthy skin | no infection |
| Staph. epidermidis | abraded skin | no infection |
| Staph. epidermidis | normal skin + intravenous line | infection |

### ■ 8.2.2 VIRULENCE/VIRULENCE FACTORS

Virulence factors can most simply be defined as the character(s) that are directly involved in the development of disease. The term **virulence** is used to grade the ability of an organism to cause disease. It is used without reference to the host but as a comparative term of the extent to which strains of the same species cause disease. The measurement of virulence is made by comparing the numbers of organisms necessary to cause disease in a suitable model. Often this involves the use of animals but wherever possible *in vitro* models are used (e.g. tissue culture systems). Figure 8.1 shows an example of an experiment comparing the virulence of two strains of the same species. Strain 2 is more virulent than strain 1 because the lethal dose 50 ($LD_{50}$) is lower for strain 2 ($2 \times 10^4$ organisms/animal) than strain 1 ($3 \times 10^6$ organisms/animal).

If the ability of an organism to cause disease resides in its possession of one or more virulence factors, it follows that the removal of this character results in a strain/mutant that has a reduced ability to cause disease. If you delete an enzyme in a central metabolic pathway that is necessary for the growth of the organism, e.g. an amino acid auxotroph, the result may be that the organism is unable to grow at all or does so at a very reduced rate compared to the wild type strain. The auxotroph is unable to cause disease in the experimental model. Does this mean that the particular enzyme is a virulence factor? To get around such awkward issues of definitions, it is necessary to distinguish between characters that are involved in basic cell physiology (as reflected in growth rate of an organism), called 'housekeeping' characters, and those that determine virulence

• **Figure 8.1** Estimation of lethal dose 50 per cent ($LD_{50}$) for two strains of the same bacterial species

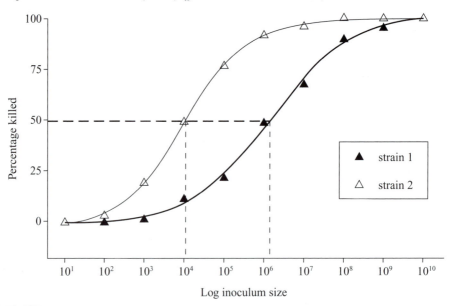

Log inoculum size

---

### ■ BOX 8.1 WHY MEASURE LD₅₀?

The slope is greatest at the mid point, reflecting the biggest increases in the proportion of animals dying per dose. This means that the difference between the $LD_{30}$ and $LD_{70}$ will be reasonably close to that of the $LD_{50}$, but values of the $LD_{90}$ may be indistinguishable between $10^7$ and $10^{10}$ organisms.

but *not* growth of the organism. Thus loss of virulence factors should not reduce the general fitness of the organism for growth. With the molecular biology skills now available to researchers, a twentieth-century set of Koch's postulates have been proposed to examine the role of individual genes and their products as virulence factors. The so-called **Koch's postulates for genes** are:

- the gene or product should be found in the strain causing disease but not in avirulent strains,
- disruption of the gene will result in reduced virulence,
- introduction of the cloned gene into an avirulent strain will increase its virulence,
- the gene is transcribed during infection,
- specific immunity to the product is protective.

It is argued that the term 'virulence factors' should only apply to those microbial products that directly interact with the host rather than any product that affects the ability of an organism to damage the host. In this way any essential microbial metabolite or product essential for general microbial growth can be excluded as virulence determinants.

To summarise thus far, virulence is a quantitative estimate whereas pathogenicity is generally used as a qualitative term. *Staph. aureus*, strain 1 may cause disease in rabbits more effectively than *Staph. aureus*, strain 2. Therefore, strain 1 is more virulent than strain 2. This latter comparison is based on the differing abilities of the organism and not the host. Descriptions of pathogenicity (e.g. Table 8.1) should take the effect of the host into account. Organism X may cause disease in mice but not rats. Organism X is therefore pathogenic for mice but non-pathogenic for rats. Likewise, Table 8.2 examines the effect of two different organisms in the same host but this time with differing circumstances. We could conclude that *Staph. aureus* was more pathogenic than *Staph. epidermidis*. An organism that is completely non-pathogenic for humans would not be able to cause an infection even when the host defences are breached, e.g. the presence of an IV line. For details of the role of the host immunity in the interplay between pathogenic organisms and humans, the reader is referred to an accompanying text in the series (*Infection and Immunity*, Davies *et al.*).

### ■ 8.2.3 INFECTIVITY

In Chapter 7 we saw that it is important to distinguish between the ability of an organism to cause disease and the ability to infect. Infection is not the same as disease. It can occur in the absence of disease and, indeed, many infections only result in a fraction of the total number infected becoming ill or showing any signs of disease. This may be memorised as:

$$\text{Infection} + \text{signs \& symptoms} = \text{disease}$$

We can distinguish between the organisms that infect humans but just become part of the normal flora of man and those that result in illness. Infections describe an acute and transient process that is progressive, whereas when organisms infect a host but become members of the normal flora in stable association without progression to pathology, the process is called **colonisation**. The term 'infectious dose' is familiar to most people and is a measure of the number of organisms required to cause an infection/disease. It can be measured for any particular organism in the same way as shown in Figure 8.1 but using different parameters. In determining infectivity, the end point is the number of organisms per animal rather than the number of animals that die. Having chosen a suitable time point, samples are taken from the animals and the number of organisms from each animal is counted. The results are plotted on a semi-log graph and the 50 per cent values are estimated to yield the infectious dose 50 ($ID_{50}$).

### ■ 8.3.2 STEP 2. INVADE

Invasion may simply reflect the attempts of a pathogen to exploit the nutritional resources that are held on the other side of the plasma membrane. Additionally, invasion may be the means by which the organism can reside for a longer period in the host. The obstacles to invasion are predominantly the skin and the epithelia lining the mucous membranes. Skin appears to be a more difficult route to penetrate unaided, therefore epithelia are the usual route of entry.

The primary function of epithelia is to provide a selective barrier between the host and the external environment. The uptake of nutrients, water and oxygen is required but the entry of bacteria is not. Bacteria move (**translocate**) from surfaces to tissues beneath the epithelia either via transcellular (entering via the cell) or via paracellular routes (passing through the cell to cell junctions). Bacteria may cross via a paracellular route by first entering mobile cells such as macrophages. Macrophages have their own mechanisms for traversing epithelial monolayers as part of their normal role of immunological surveillance. A distinction can be made between bacteria that target phagocytic cells such as macrophages and neutrophils and those that enter non-phagocytic cells such as epithelia. The research on enteric pathogens has yielded many insights into the mechanisms of entry in epithelia. Specific adhesins called **invasins** trigger the host cell membrane to engulf the attached bacterium such that the organism enters without having to disrupt and thereby kill the host cell. It is clear that such a tidy mechanism avoids the release of compounds that will attract the host immune cells. One of the most surprising elements in the strategies of bacterial pathogens is the extent to which the bacterium controls the situation. It is now evident that bacteria such as salmonellae secrete regulatory molecules that control events in the host cell using the various protein secretion systems. Many bacterial virulence factors are proteins that need to be excreted so as to exert their action on the host cells. Examples include haemolysins, proteases, lipases and nucleases, along with the classic excreted exotoxins. Of the four currently recognised secretion systems, type III is of particular interest in that it mediates the transfer of a bacterial signalling molecule into the host cell. Viewed as a syringe with which to inject the molecules into the recipient host cell, it requires physical contact (contact dependent or contact mediated) with the host cell for it to work. Examples of the activity of these signalling molecules (effector proteins) are the Sip proteins injected by salmonellae, which cause rearrangement of the entero-cyte membrane and a subsequent ruffling appearance. These events facilitate the uptake of the salmonella into the host cell.

### ■ 8.3.3 STEP 3. MULTIPLY

Little is known about the multiplication of bacteria in human or animal tissues during infections. Obviously, the conditions in which bacteria are cultured in the laboratory do not accurately mimic those inside cells or tissues. One of the recognised differences is the concentration of free (available) iron. Unlike culture media, animal tissues hold iron at very low concentrations ($10^{-9}$M) by use of iron-binding protein carriers. In the extracellular fluids transferrin and lactoferrin are used, whereas inside mammalian cells iron that is not complexed in specific proteins (e.g. heme proteins) is held in ferritin. For bacteria to multiply in human tissues the production of siderophores to extract iron from the iron-binding proteins is an essential and rate-limiting step (see Chapter 2 for details of iron limitation).

Other variables include gaseous conditions ($O_2$, $CO_2$), redox potential, pH and osmolarity, all of which will change depending on the location within the body. This

Havin;
immune 1
of invadir
attempt t
escaping
by which
viruses, s

**8.3.5.1**
Consistin
tection f
numerou
*influenzae*
three org
normal h
tection a,
which pe

**8.3.5.2**
Accordin
proteins
'outer m
their fun
when Gr.
duction c
minal m
termed :
humoral

**8.3.5.3**
IgA has b
epithelial
defences
number c
tory IgA
*influenzae*
pathogen

**8.3.5.4**
Immunol
immune
inside ho

**8.3.5.5**
Surface s
ducts tha
and oute
or contr(
antigens
The f

must be compared with the closed conditions within a batch culture which can be predicted (and measured). Further complications in the measurement of growth rates *in vivo* and *in vitro* include the problems of bacterial lag times (which may vary between site within the host) and the numbers of organisms killed by the host defences (e.g. the numbers multiplying may balance the numbers killed). Such problems mean that measured doubling times in culture media are likely to be artificial. Furthermore, it is often assumed that increasing growth rate *in vivo* equates with virulence and that rapid multiplication *in vivo* is preferred. However, the slow-growing *Mycobacterium tuberculosis* provides an example of an alternative plan.

The ability of the pathogen to utilise nutrients that are particularly abundant at the site of infection makes sense, even if specific examples are few. The ability to break down urea by urease in the urinary tract would be a good example, but unfortunately the most frequent cause of human urinary tract infections is *Esch. coli*, which does not possess ureases. *Proteus spp.*, which do produce urease, are less common. Obtaining sufficient nutrition is an important limitation on growth *in vivo* and of particular interest is the uptake of nutrients of intracellular pathogens. Virulence factors that are used to obtain nutrients from sites have been termed **pabulins**. One proposed mechanism for obtaining scarce and limiting nutrients is to lyse the host cell with cytolytic toxins. The regulation of the diphtheria toxin by concentrations of free iron is a strong argument in favour of the centrality of nutrient acquisition driving the virulence of any pathogen.

The three cardinal features of pathogenic bacteria discussed (attach, invade and multiply) are those considered to be the minimum needed to create an infection. Additional characteristics can be found in some but not all pathogens. These include:

- the ability to spread within the host,
- the ability to persist within the host, and
- the means with which to exit the host.

### ■ 8.3.4 STEP 4. SPREAD WITHIN THE HOST

The variety of clinical infections outlined in Chapter 7 tells us that not all infections follow identical paths. Options open to pathogenic bacteria are to multiply only at the initial site of entry (*Corynebacterium diphtheriae*) or multiply at a site distinct from the site of entry. Alternatively, bacteria may spread at both the entry site and elsewhere. The extent to which the infection spreads gives rise to the terms 'local' or 'generalised' infections. The spread to other organs/tissues may simply be passive, i.e. the infection spreads into adjacent tissues (e.g. staphylococcal abscess) or spills over into the bloodstream, and the organisms are carried to other sites in the host. In such cases the infections in the distant sites will only be seen in a proportion of cases and therefore are considered as complications. In systemic infections such as typhoid fever, the organism (*Salmonella typhi*) takes a route through the host that encompasses several distinct organs/tissues in the host. The ordered sequential multiplication in different organs is what defines the disease in humans. The attachment and entry via the M cells in the intestine is followed by migration in macrophages to the lymph nodes and subsequently to the spleen, liver and bone marrow. From here the organisms multiply and then subsequently invade the bloodstream, the intestine and gallbladder. This staged course of events taken by the organism in the course of its life in humans could not occur by accidental spread of the organism through the host.

The mechanisms of spread throughout the host can vary. The routes are as follows:

## RECOMMENDED READING

Alouf, J.E. and Freer, J.H. (eds) (1999) *The Comprehensive Sourcebook of Bacterial Protein Toxins*, 2nd edition, Academic Press, London, UK.

Davies, D.H., Halablab, M.A., Clarke, J., Cox, F.E.G. and Young, T.W.K. (1999) *Infection and Immunity*, Taylor & Francis, London, UK.

Finlay, B.B. and Falkow, S. (1989) Common themes in microbial pathogenicity. *Microbiol. Rev.* 53, 210–30.

Finlay, B.B. and Falkow, S. (1997) Common themes in microbial pathogenicity revisited. *Microbiol. Mol. Biol. Rev.* 61, 136–69.

Mims, C.A., Nash, A. and Stephen, J. (2000) *Mims' Pathogenesis of Infectious Disease*, 5th edition, Academic Press, London, UK.

Orndorff, P.E. Bacterial pathogenicity, in Balows, A., Truper, H.G., Dworkin, M., Harder, W. and Schleifer, K.-H. (1991) *The Prokaryotes. A Handbook on the Biology of Bacteria*, 2nd edition, Springer, Berlin.

Salyers, A.A. and Whitt, D.D. (1994) *Bacterial Pathogenesis: a Molecular Approach*, ASM Press, Washington, USA.

Schlaecter, M., Engelberg, N.C., Eisenstein, B.I. and Medoff, G. (eds) (1998) *Mechanisms of Microbial Disease*, 3rd edition, Lippincott, Williams & Wilkins, Baltimore, USA,

## REVIEW QUESTIONS

*Question 8.1*    What infectious diseases do not follow the six steps followed by a pathogen?

*Question 8.2*    If rapid replication rate is a key feature of bacterial pathogens, is *Mycobacterium tuberculosis* a successful pathogen?

*Question 8.3*    Compare and contrast bacterial exotoxins and endotoxin.

*Question 8.4*    What are the problems with the original Koch's postulates?

At a conservative estimate, there are at least thirty different viruses from fifteen taxonomic families that are common causes of human illness. This number does not take into consideration the serological types that exist in many of these viruses. So the common cold we are taking as one virus, whereas there are over eighty serological types of rhinovirus. As every human will be infected by viruses in their lifetime, we must concede that viruses are efficient parasites of humans. This chapter outlines some of the features of viruses that contribute to their success as human pathogens.

## ■ 9.1 TYPES OF CLINICAL INFECTION

The four broad types of viral infections recognised in humans are listed in Table 9.1, and the relative time course for each is shown in Figure 9.1.

Because it is only those people with symptoms who come to our attention, it is easy to overlook the fact that most viral infections are asymptomatic. For the people who demonstrate clinical illness, the severity will range between mild and severe. The proportion of susceptible people who develop illness from those who are infected is called the **attack rate** (Table 9.2), although it is not a true 'rate' and gives no indication of the total number of people infected.

The route of transmission does not fully account for the variation in attack rate between viruses, although clearly direct inoculation by parenteral transmission is likely

**Table 9.1 Types of viral infection**

| Duration | Extent | Examples |
|---|---|---|
| Acute | Local | rhinovirus: the common cold |
| | Systemic | measles |
| Persistent | | chronic hepatitis B |
| | | Epstein Barr virus |
| Latent | | Herpes simplex: cold sore |
| | | Varicella zoster: shingles |
| Slow | | Bovine spongieform encephalopathy |

children with little or no pathology and then persist for life. This lifetime of infection will optimise the chances of transmission, preferably to other children. Persistent infections, by definition, require that the virus is maintained within the host cell and thus seeks to avoid killing the cell (is not cytolytic). Integration of the viral nucleic acid into the host chromosome is an appropriate means to help promote long-term survival in the host and avoid contact with the immune system which will also seek to destroy the infected cells. Whilst persistent virus infections are mostly DNA viruses such as Herpes and papillomaviruses, retroviruses will insert themselves as proviral DNA into the host genome following reverse transcription from RNA and many retrovirus infections are persistent.

The differences in acute and persistent infections have consequences for antiviral treatments. Specific antiviral chemotherapy will only have significant effects on persistent infections where there is continued production of virus. In acute infections, the virus is produced in greatest numbers just prior to the development of symptoms. By the time the patient has sought help, the virus is no longer in full manufacture.

Other differences result from the properties of DNA and RNA viruses. You may recall from Chapter 3 that viruses contain either DNA or RNA and not both. RNA viruses have a reduced gene capacity compared with DNA viruses. DNA viruses often possess sufficient genetic space (several hundred genes) such that roughly 50 per cent code for proteins that can interfere with the host immunity (see below, p. 219). RNA viruses do not have sufficient gene space and utilise genetic variation in order to generate new variants of virus.

## ■ 9.3 STAGES IN HUMAN INFECTIONS

> The key stages in the process of an infection are:
>
> - entry into host,
> - primary replication,
> - spread within host,
> - exit from host,
> - evade host response.

## ■ 9.3.1 ENTRY INTO HOST

The routes of transmission will depend on the site of the body that represent the exit site for the infectious virus. The exit of virus from the body can occur via several vehicles: urine, faeces, semen, saliva, blood, milk, skin, squamous epithelia. Hepatitis B virus is released into the bloodstream from the liver and the transfer of infected blood is the major vehicle for transmission. Examples include the sharing of needles amongst drug abusers and the accidental transmission following blood transfusion. With hepatitis B the transmission is parenteral and therefore the virus cannot be transmitted by sneezing or touching. Note how the site of multiplication (the liver) need not be the infectious site (blood). Conversely, rhinoviruses, the aetiological agents of the common cold, multiply in the upper respiratory tract epithelia and are transmitted via the secretions from the runny nose.

There are several ways of listing routes of transmission of infectious agents (as given in Chapter 7). The traditional list is as follows:

- aerosol,
- faeco-oral,
- parenteral (including vertical transmission),
- sexually,
- direct contact.

The routes of horizontal transmission can be related to the sites in the human body that represent the major sites of entry:

- skin,
- gastrointestinal tract,
- respiratory tract,
- genito-urinary tract.

This list can be sensibly reduced by grouping the last three into one to yield three key routes of entry for a virus into a new human host:

- **Entry through mucous membranes**: respiratory, genito-urinary and gastrointestinal tracts and the conjunctiva.
- **Entry through skin**: insect bites (e.g. yellow fever virus); animal bites (e.g. rabies virus); skin to skin (e.g. papillomavirus (warts, veruccas), Orf virus).
- **Parenteral transmission**: vertical transmission, direct inoculation into blood (needle, insect or animal bite) or blood transfer through trauma (including sexual transmission).

Vertical transmission has clear advantages in that the foetus has the same blood supply as the mother and has no defences against virus attack other than those present in the mother. One of the drawbacks of vertical transmission is the risk of damaging the developing child *in utero*, i.e. foetal death and abortion. A less extreme outcome is teratogenesis (see Box 9.1). Vertical transmission also includes transmission via the germ line (i.e. through infected sperm or ovaries).

The properties of the virus will also influence those routes that are likely to be successful for the virus. Viruses that cause gastroenteritis will need to resist exposure to

---

■ **BOX 9.1 VIRAL TERATOGENESIS**

**Teratogenesis** is the ability to cause defects in the foetus. The rubella virus is one of very few viruses that are able to cause congenital malformations in the developing foetus. The vertical transmission of the virus occurs simply through the unfortunate timing whereby the mother becomes infected with the virus for the first time (i.e. is susceptible) in the first three months of pregnancy. When the mother becomes viraemic the virus is able to traverse the defences of the placenta and infect the growing child *in utero*. Up to 30 per cent of babies infected in the first trimester are born with malformations such as neuronal, eye and heart defects. Infections acquired in the last two trimesters rarely result in any defects. Congenital rubella occurs because the virus infects cells in a key stage of development. Severe infection of the foetus will be unsustainable and lead to stillbirth or abortion. Children born with congenital rubella are likely to represent those that have been able to repair and develop sufficiently, in spite of the viral damage.

**Table 9.5 Transforming viruses**

| Virus | Family | Viral nucleic acid |
| --- | --- | --- |
| SV40 | Papovavirus | DNA |
| Human papillomavirus* | Papovavirus | DNA |
| Epstein Barr virus* | Herpes virus | DNA |
| Rous sarcoma virus | Retrovirus | RNA |
| Human T-cell lymphotrophic virus* | Retrovirus | RNA |

*Viruses that infect humans.

The use of the term 'transformation' applies to virus-infected eukaryotic cells and should be distinguished from the transformation of bacterial cells by the uptake of naked DNA (as described in Chapter 1).

and retroviruses (the DNA is synthesised from the viral RNA using reverse transcriptase and subsequently incorporated into the host genome). It is important to place these events in perspective. Transformation is a rare event since only certain viruses can transform human cells (Table 9.5) and the frequency is low (e.g. approximately 1 in every $10^5$ cells infected with SV40 virus). Finally, cells that are transformed are not necessarily neoplastic (tumour forming) because transformation is only one stage of many in the process where a eukaryotic cell becomes neoplastic and it does not follow that a transformed cell will become malignant.

### ■ 9.5 THE RESPONSE OF HOST CELLS TO VIRUS INFECTION

An immune response is necessary for the prevention of uncontrolled proliferation of virus. Viruses will need to suppress the immune response sufficient to permit adequate replication and transmission. Complete suppression is unnecessary and would be costly in terms of viral resources and would allow the proliferation of other micro-organisms.

The balance that exists between mounting a vigorous immune response that is damaging to the host tissues and not being able to contain the virus can be illustrated in experiments in people suffering from the common cold. The general miserable symptoms were reduced by patients who took anti-inflammatory pain killers *but* the group who took nothing had a shorter period of illness at the expense of feeling worse.

It is said that viruses do not produce typical toxins as seen in toxigenic bacteria. This might reflect the overall limits in our knowledge of viral genomes. Certainly it has become clear that viruses do code for proteins that are potentially harmful. Several of the viruses that are linked with human cancers produce proteins that act to prevent the host cell from undergoing **apoptosis** (programmed cell suicide). Apoptosis should be contrasted with necrosis. It is not possible to summarise the toxicity of viruses into one dominant effect because of the diverse strategies employed by the viruses to successfully replicate and be transmitted.

The mammalian cell has developed several mechanisms to try to limit the virus from overwhelming the host. Whilst we all recognise the symptoms of a developing cold or other viral infection, the extent to which this directly represents the toxicity of the viral infection or our reaction to an invader is not clear. The host response to infection at the cellular level is an appropriate place to examine the fight back.

The human host responds to all infectious agents with cell-mediated and humoral responses. Against intracellular pathogens, cell-mediated responses are generally more important since, although antibodies can mop up free virus in body fluids such as the bloodstream, they cannot penetrate cells. For antiviral immunity, one important group of polypeptides are the **chemokines** and **cytokines**. These small molecular weight proteins regulate the interaction between immediate immune response (the **innate immunity**) and the adaptive immune response that confers **specific immunity**. Chemokines and cytokines have homeostatic and proinflammatory functions. The former concerns the maintenance of haematopoietic system (lymphoid cell lineages from the bone marrow) and the latter is involved in the recruitment of leukocytes from the circulation to the sites of infection. Chemokines/cytokines act on cells through

binding to appropriate cell surface receptors triggering the synthesis of antiviral proteins. Chemokine receptors have proved to be important in viral infections through
the discovery that the HIV binds to a particular chemokine receptor on lymphocytes.
The people with the rare congenital deficiency in this receptor are resistant to infection
with HIV. Cytokines are released from activated immune cells in response to infection
and they exert antiviral activity. There are increasing numbers of cytokines being identified but the key antiviral cytokines are **interferon-alpha, -beta, -gamma** and
**tumour necrosis factor**. Cytokines bind to specific receptors on the appropriate
host cells (typically macrophages and lymphocytes) which activate signalling pathways
that, in turn, activate various intracellular antiviral events. Many of the mechanisms are
as yet unknown but the inhibition of the replication of the virus is the net effect. The
antiviral mechanisms activated by cytokines can result in the lysis of the infected cell or
more subtle interference of viral replication without killing the infected host cell.
Destroying infected cells may be too damaging when dealing with cells that cannot
regenerate, for example nerves or in massive infections of a vital organ (e.g. extensive
liver damage by hepatitis viruses) and a non-cytolytic, non-cytopathic mechanism is
required.

## ■ 9.5.1 PRODUCTION OF INTERFERON

It was demonstrated in the 1950s that the culture medium bathing virus infected cell
lines could inhibit the multiplication of viruses in separate flasks. The active substances,
termed **interferons**, are cytokines (as discussed above) and are produced within hours
of virus infection to limit the spread of virus in the host whilst specific immune
responses are developing. Most cell types can synthesise interferon but they can be
grouped into three broad types according to the cell type that manufactures them.

- IFN-$\alpha$ (alpha): leucocyte interferon,
- IFN-$\beta$ (beta): fibroblast interferon,
- IFN-$\delta$ (delta): immune interferon (activated T-lymphocytes and NK cells).

Double-stranded RNA is the most potent stimulus for the production and release of
interferon alpha and beta which act to prevent or limit the surrounding cells from
becoming infected. Interferons are prophylactic rather than curative. The actions of
interferons are somewhat varied in that they can modulate the activity of immune cells
as well as induce resistance to viral infection. Two important antiviral processes
induced by interferon in virus-infected cells are:

- the inhibition of viral-induced protein synthesis, and
- degradation of viral mRNA and rRNA.

The central player in cells stimulated by IFN is **protein kinase R** (PKR). The letter R
is taken from the viral ds*R*NA that activates the transcription of this protein kinase.
PKR produced by cells stimulated by IFN inhibits viral driven protein synthesis by
binding to the double-stranded RNA. PKR will also trigger apoptosis, presumably as a
last resort following uncontrolled viral replication.

## ■ 9.5.2 TRIGGERING APOPTOSIS

Controlled cell death (**apoptosis**) is a mechanism by which multicellular organisms
remove unwanted cells either during development (wonderfully illustrated by the loss

of the tail in the developing tadpole) or respond to genomic abnormalities (pre-cancerous changes). Apoptosis is also of great value as a defence mechanism against virus infection in cells. Apoptosis, in contrast to necrosis, is a tightly controlled process that leaves no mess and causes no damage to surrounding cells. As with host cell nucleases, viruses have acquired mechanisms with which to counteract the triggering of apoptosis.

## ■ 9.6 VIRUS-INDUCED TISSUE DAMAGE: VIRUS OR HOST?

To what extent does the replication of the virus in the tissue cause the disease? The varied outcomes following a virus infecting a cell are given above (p. 213). Some of these help explain why virus infections can cause cell damage. For example, lytic infections will result in the loss of function of the infected cells. If sufficient numbers of cells are damaged in this way then the function of the organ may be compromised. Rotaviruses cause diarrhoea in humans and animals. The infected cells of the villi of the ileum are shed resulting in the loss of absorptive surface area contributing to the diarrhoea. The diarrhoea is a direct consequence of the viral damage in the enterocytes.

The common symptoms of many viral infections, myalgia and headache, are usually indirect consequences of the host immune response (often cytokines such as tumour necrosis factors) rather than direct viral replication in the muscles and brain.

Other examples of tissue damage are indirect. In true clinical poliomyelitis the virus damages the nerve cells that serve the muscle cells of the limbs. The damage to the nerve cells by the virus causes the muscle to atrophy. Virus cannot be found in the affected muscles themselves. Furthermore, poliomyelitis virus normally only multiplies within the enterocytes of the small intestine and they show no morphological alterations. This serves to remind us that the CPE observed in the cell lines used to grow viruses in the laboratory are not necessarily reflected in the natural host cells.

Different infection strategies are employed by different viruses. The extremes are an acute, hit-and-run approach or a chronic persistent type of infection. Viruses choosing the former option will have less concern for the effects of rapid viral multiplication on the host. Such viruses will tend to suppress or inactivate acute (innate) mechanisms of host defence so as to gain time for the rapid multiplication of new virus. If the host is fatally damaged by the infection, as long as sufficient virus has been manufactured such that the infection is transmitted to new hosts, then the death of the host is of no concern. Aggressive virus multiplication will probably result in protective immunity such that, if the host recovers, he or she will not be available for the virus to reinfect. Chronic infections will need to adopt those strategies that deal with the longer term problems of specific immunity, i.e. the development of specific antibodies and cell-mediated cytotoxicity.

The reaction of the host to intracellular parasites, viruses in particular, will often result in tissue damage. Immunopathological damage will cause tissue pathology by at least two mechanisms.

### ■ 9.6.1 IMMUNE CELL-MEDIATED CYTOTOXICITY

In tissues with rapid turnover, virus-infected cells that are lost may be replaced (e.g. intestinal and skin epithelia). In non-replicating tissues like the heart and nerve tissue virus-infected cells cannot be lost so readily without possible functional deficit. If, therefore, the immune response damages these cells, the organ suffers from impaired function. Hepatitis viruses do not cause lytic infections of the hepatocyte but instead evoke a lymphocytic cell inflammation. This cell-mediated host response is what damages the liver tissue with resulting impairment of function.

### ■ 9.6.2 IMMUNE COMPLEX REACTIONS

Immune complex reactions are typical of persistent infections. The antigen–antibody complexes can get stuck ('deposit' themselves) in arteries to cause vasculitis (which manifests itself as a skin rash) or in the basement membrane of the glomeruli to cause kidney damage. Both are seen with chronic hepatitis B infections.

# ■ 9.7 HOW VIRUSES DEAL WITH HOST IMMUNE RESPONSES

The host immune response to viruses is both humoral and cell-mediated. Circulating antibody will neutralise extracellular virus in the body fluids (including mucous membranes) and the cell-mediated arm will attack cells that express viral proteins. Previous infection or vaccination will be the two relevant factors that dictate whether a person is able to swiftly neutralise an infection or not. In light of these protective mechanisms, viruses have developed strategies to try to minimise their contact with the immune response of the host. Viruses have adopted various strategies to out-manoeuvre the host, in particular to evade or suppress the immune response that attempts to develop antiviral products.

It is useful to consider the broader strategies before looking at cellular mechanisms.

- **Evasion of immune response.** Viruses can replicate in tissues that are relatively protected from surveillance by immune cells (e.g. brain, dermis). Alternatively, they can avoid extracellular stages in the course of infection. Fusion of infected cells, budding within cell vesicles and replication within lymphocytes all avoid exposure to neutralising antibodies.
- **Suppression of immune response.** Some virus infections infect the immune cells that mount the immune response in order to suppress them. Strategically this makes good sense; hence, many viruses infect lymphocytes, e.g. B lymphocytes are infected by EBV and T lymphocytes are infected by HIV. (It is interesting to note that intracellular bacteria tend to infect macrophages, e.g. *Mycobacterium tuberculosis*.)

Virus infections of the neutrophils appear to be either unusual or of little value to the virus, possibly because neutrophils do not live long enough for adequate viral replication. Neutrophils live for only 12–18 hours, whereas lymphocytes and macrophages have long lifespans which make them worthwhile targets.

# ■ 9.7.1 CELLULAR MECHANISMS BY WHICH VIRUSES MODULATE THE HOST IMMUNE RESPONSES

As we have seen, the first line response (innate immunity) of the host to viral infections is dominated by the actions of the chemokines called interferons. Interferon will need to contain viral infection until the specific immunity has developed. Many viruses have developed the ability to block the actions of interferon-based responses by producing proteins that interfere at various points in the action of the different chemokines/cytokines. It is obvious that viruses need to thwart the early attempts of the immune system to kill them. In order to multiply and then exit the host before the specific immunity develops, the virus must suppress the innate immunity for as long as possible. It has been pointed out that total suppression of the immune response is not a feature of viral infections. Despite being costly in terms of genome space for the virus, this strategy will be of benefit since it wants to multiply in the host at the expense of other invading parasites. Selective suppression of the innate immunity still enables the host to repel other pathogens.

The host response to virus infection will generate extracellular signals (cytokines) and these will bind to cell receptors and trigger intracellular events that will aim to inhibit viral replication. It follows that viruses can inhibit both the extracellular and intracellular signalling events. The following list illustrates how viruses have attempted to modulate various points of the immune response.

- **Minimise recognition by the host.** For viruses that cause systemic infections, the first entry of virus into the bloodstream during a viraemic phase will attract the attention of Complement components. Certain viruses produce molecular mimics (**homologues**) of Complement components which will block the Complement cascade attacking the free virion. These homologues are termed **viroceptors**. Mimicry is also used by enveloped viruses which include host proteins in the viral envelope to mask their recognition as 'foreign'. Virus-infected cells also secrete proteins that suppress their recognition by Complement, thereby protecting the manufacture of new virus.
- **Inhibit production of interferon.** Interferon production is an important early step in the defence against virus infections. Viruses need to interfere with the action of interferon to establish an infection. One way to counter the action of interferon, is for viruses to secrete molecular mimics of IFN receptors. In this way IFN binds to the competing mimic receptor rather than real receptor on the virus-infected cell. As molecular mimics of receptors, they are called **viroceptors**. Other viral proteins have been found to target the cellular signalling pathways that occur following binding of IFN to the cell receptor.
- **Modulation of cytokine action.** The mechanisms that viruses target when interfering with the action of cytokines, most notably interferon, may be at any point in the communication network employed by cytokines:

**interfering with cytokine production by cells**
**inhibiting cytokine binding to cells** and
**interfering with the intracellular action of cytokines**

The action of viruses on cytokines, however, is not always inhibitory. Some viruses actually recruit cytokines that promote cellular proliferation. In this way the net production of the virus will be enhanced.

- **Prevent apoptosis.** Apoptosis, programmed cell suicide, occurs only if a cell receives a signal to proceed. That signal may be triggered by the presence of a virus. In general, viruses try to prevent the cell from carrying out this self-destruction and numerous examples exist of viruses producing proteins that inhibit apoptosis. Apoptosis of eukaryotic cells can be triggered following activation of proteins such as p53 and ICE-like proteases. Certain viruses produce proteins that inhibit the functioning of these two triggers.

In order to produce viroceptors and other homologues, the virus must possess sufficient gene space. This is only possible in the larger DNA viruses such as Herpes and pox viruses. RNA viruses do not have the genome size to carry the necessary protein coding. To compensate, RNA viruses will more often choose to outwit the host by exploiting the genetic plasticity inherent in replicating RNA. One of the mechanisms for creating genetic variability and thereby generating time whilst the host develops specific immunity is **antigenic variation**. The host has spent a week manufacturing immunity against the original strain of virus but antigenic variation (Figure 9.2) creates a new strain against which the host is immunologically naïve.

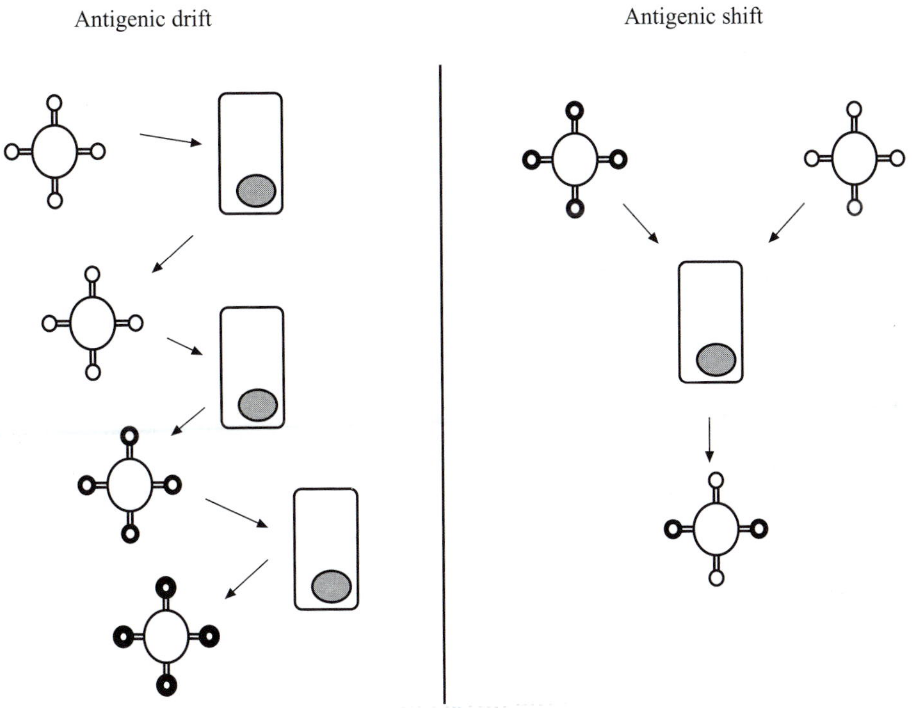

• **Figure 9.2** Antigenic drift and antigenic shift. Antigenic drift (left) shows how the random mutations of a single base during replication of the viral RNA results in a new amino acid within a surface protein (haemaglutinin or neuraminidase in influenza A virus). The modified antigen is recognised partially by antibodies generated against the original version. Antigenic shift (right), however, represents a more substantial gene rearrangement. In this case the protein is substantially altered through reassortment from a cell co-infected by two distinct strains of virus such that there is no protection conferred from antibodies against previous types of the antigen. Epidemics and pandemics may follow antigenic shift as the population is effectively completely susceptible again

Antigenic drift                                                           Antigenic shift

### 9.7.1.1 Antigenic variation

Antigenic variation occurs both during repeated infections over the lifetime of the host (influenza viruses) and during the course of a single infection of the host (HIV virus). As antigenic variation means that some types will be more immunogenic than others, the strategy is one of some risk but selection will promote those variants that are most suited to infect and transmit themselves through a population.

Viruses exploit three mechanisms for creating genetic diversity:

*   **spontaneous mutation,**
*   **recombination,**
*   **reassortment.**

**Spontaneous mutation.** In the course of replicating RNA mutations arise with a much higher frequency than in DNA replication. The spontaneous mutation rate is close to one mutation every round of replication of the genome, the highest rate in all living organisms when compared to rates of <0.01 for DNA viruses and bacteria. The major cause of this hypermutability is the poor accuracy (low fidelity) of the RNA-dependent RNA polymerase which has only a nominal ability to check and correct mistakes in the base sequences. Whilst mutations may be good (increase fitness) or bad (decrease fitness) other features of viral infections serve to increase the rate at which new viruses evolve. This includes viral replication itself, which yields large numbers of new virus in a short period of time. Within a human host, selection will eliminate those mutants that are less well suited, with the result that only a tiny fraction of the new viruses will be generated within the course of an infection. The total number of viruses manufactured in a host will be a collection of closely related genomes, dominated by the 'wild type' genome but including a range of mutants of rapidly diminishing proportions. This distribution of non-identical but related viruses is called a **quasispecies**. The variation in the genome that leads to minor sequence base changes are exemplified as **antigenic drift** in influenza viruses (Figure 9.2). The antigenic variation occurs in the viral haemagglutinin, a protein that projects from the virion and mediates binding to red blood cells. Antigenic drift results in the reduced affinity of host-neutralising antibodies.

**Recombination.** Viral recombination occurs when a cell is infected with two different RNA viruses and the RNA produced from each virus undergoes strand switching. Exchanges of genome fragments under these circumstances is not strictly controlled and the products may be deleterious mutants. Viral recombination is able to introduce more genetic changes than single mutations, but for entire genes to mix a virus needs genetic reassortment.

**Reassortment.** Influenza virus has a segmented genome and when two strains infect the same cell, the reassortment of segments from the two genomes are packaged in new combinations to yield a new strain of virus. Reassortment is responsible for the phenomenon of **antigenic shift** (Figure 9.2) where both **haemagglutinin** and **neuraminidase** proteins of influenza virus are antigenically distinct from the originals. The loss of protection of previous neutralising antibodies partly explains why epidemic influenza occurs after antigenic switching.

The importance of viral genetic variation lies in the ability it gives RNA viruses to repeatedly infect the same host, as exemplified by annual rounds of new influenza virus strains. The implications of viral quasispecies are witnessed in AIDS patients where microbiologists have had to introduce multiple antiviral drug combinations for HIV treatment so as to minimise resistance developing from the selection for resistant mutants by one drug. Similarly, there are implications for vaccine design if new variants are readily generated, leaving the immunised population susceptible.

## ■ 9.8 VIRUSES AND CANCER

It may surprise people to hear that 10–15 per cent of all cancers across the globe can be attributed to viruses. This figure is likely to increase as research uncovers more viral genes within the tumour cells. One small benefit that has come about from the pandemic of AIDS has been the discovery that a new herpes virus, human herpes virus-8 (HHV-8) has been identified to play a role in the development of Kaposi's sarcoma. The importance of understanding the role of viruses in the aetiology of cancer cannot be overemphasised. Studies of viruses have led to many of the critical discoveries in the workings of eukaryote cell biology: gene mutations in bacteriophages, cellular oncogenes, tumour suppressor genes, bacterial restriction enzymes and viral reverse transcriptase

All of these have been uncovered through the study of viruses. The study of such fundamental mechanisms of cell biology provide new targets for the design of novel anticancer and antiviral drugs. Perhaps the most powerful implication from studying infectious links with cancer is the remarkable finding that vaccination can be used to prevent cancer. Vaccination against hepatitis B has reduced morbidity and mortality of liver cancer in Taiwan.

Cancer is the uncontrolled proliferation of cells resulting from the accumulation of mutations in genes that regulate the cell cycle. Normally, cell proliferation is tightly controlled such that cell division is matched by cell loss in order to maintain the size of an organ (e.g. the liver). The cells are able to respond in a balanced manner to the signals that promote or inhibit growth. In cancer the regulatory gene products malfunction and the cell divides continuously. Carcinogenesis is a multistep process: a series of events (up to nine in the case of colonic cancer) need to occur before the cancerous properties are fully developed. Cancer is also multifactorial in origin: several different triggers are required to cause the mutations, of which viral infection may be just one.

A number of viruses have been shown to directly cause cancer in animals. Rous in 1910 induced sarcoma in chickens by injecting filterable extracts from diseased animals. The filterable agent turned out to be a retrovirus and numerous retroviruses have been found subsequently to cause animal tumours. Such a direct causal relationship has not been so easy to establish in humans. Some of the viruses that have been linked with human cancer are as follows.

- **Epstein Barr virus and Burkitt's lymphoma.** The first association of a virus with a human cancer was in 1964 with the direct observation of Epstein Barr virus particles by electron microscopy in cell cultures from humans with Burkitt's lymphoma. This cancer is geographically limited to the malaria belt of Africa. Hybridisation with viral nucleic acid was also found in tumour biopsies.
- **Hepatitis B and hepatocellular carcinoma.** Seroepidemiological studies in the 1970s linked the virus with hepatitis. Hepatoma is most prevalent in parts of the world where HBV is most common. Vaccination against childhood hepatoma has fallen significantly since introduction of hepatitis B vaccine. A precedent was found in animals when HBV-like viruses were found in cases of hepatitis and hepatoma in woodchucks!
- **Human papillomavirus (HPV) and cervical carcinoma.** There are over 80 antigenic types of HPV which infect skin or mucous membranes. HPV show strong tissue tropism, hence the different antigenic types are associated with particular lesions: mostly these are benign cellular proliferations such as warts, macules and papillomas of skin and mucous membranes. HPV are DNA viruses which cannot be cultured *in vitro* but, instead, require nucleic acid hybridisation for identification in tissue. Cervical carcinoma is the most common in the developing world and because it is rare in nuns and very rare in virgins compared with people of multiple sexual partners, the sexually transmitted *Herpes simplex* virus was originally suspected. However, HPV came under suspicion in 1975 following weak hybridisation in cervical tissue using a probe from HPV of warts. In 1983, HPV-16 and HPV-18 genome were isolated from cervical cancer biopsies. HPV virus genes are expressed in most malignant tissues and in the HeLa cervical cell line. A series of trials are underway in the UK to examine the effect of vaccination against HPV-16 on cervical cancer.

### ■ 9.8.1 PROOF OR EVIDENCE?

A number of general observations can be made concerning the role of viruses in cancer. Cancer develops decades after primary infection, giving ample opportunity for chemical and physical (radiation) carcinogens to act. Many of the suspect viruses (EBV, HPV, hepatitis viruses) are ubiquitous across the globe yet only a small percentage of infected people develop the associated cancer. Most people will clear the viral genome or, failing that, coexist without symptoms or disease.

## ■ 9.8.2 CELLULAR MECHANISMS OF VIRAL CARCINOGENESIS

There are two sets of genes and their products that are involved in the mechanisms by which viruses can trigger immortalisation of cells:

- tumour suppressor genes,
- oncogenes.

### 9.9.2.1 Tumour suppressor genes

Most of the cells in the human body are not dividing but quiescent (metabolically active but not dividing). They are growth-arrested and, in terms of the cell cycle, in G0/G1 phase. Throughout the cell cycle there are checkpoints during which the damaged or mutated DNA is either repaired or the cell undergoes apoptosis. Of the proteins involved in surveillance, the two most important are p53 and pRb and, because they regulate the removal or repair of potentially malignant cells, they are termed **tumour suppressor genes**. Viruses need to prevent cells from recognising the foreign viral DNA and suppressing any proliferation (viral transformation). Likewise, apoptosis must be prevented or delayed until viral replication has finished. DNA viruses will force the cell to undergo DNA synthesis by producing early viral proteins that trigger DNA replication by sending the cell from G0/G1 into S phase. This can be achieved by either inhibiting the proteins (p53 and pRb) that prevent progression into S phase or stimulating the expression of growth-promoting genes.

### 9.9.2.2 Oncogenes

Genes that cause transformation of cells are called **oncogenes**. Strictly, oncogenes are mutated forms of proto-oncogenes. Proto-oncogenes are normal cell genes that are involved in cell signalling events that regulate growth and division. Proto-oncogenes are also referred to as c-oncs to distinguish them from viral oncogenes (v-oncs). Viral oncogenes are modified from cellular oncogenes and present only in RNA viruses. DNA viruses do not posses oncogenes. For RNA viruses to exert an oncogenic effect they must have a stage in their replication cycle when they produce DNA and insert this as a provirus in the host genome. Retroviruses are such viruses (although hepatitis B virus can also do this). Oncogenic viruses that transform cells act by inserting a v-onc into the genome or by insertional activation (the viral promotor promotes the transcription of a c-onc). The mechanisms are illustrated in Figure 9.3.

## ■ 9.9 CAUTIONARY POINTS

Before one leaves with the impression that all viruses cause cancer, we should remind ourselves that the multistep nature of carcinogenesis means that a virus that has one or two of the properties described is still unlikely to induce malignant growth. For example, EBV is known to immortalise B-lymphocytes (*in vitro*) from all people, not just people in the malaria belt, thus indicating that additional events are required to lead to lymphoma. Indeed the multiple trigger hypothesis might explain geographic restriction of certain tumours where malaria, burnt salt fish, herbal snuffs and other phenomena peculiar to specific cultures act as cofactors. Some viruses can immortalise cell lines *in vitro* but have no known associated human cancer (polyoma viruses). Conversely, some viruses (e.g. HBV linked with hepatoma) do *not* immortalise cell lines.

Whilst immunosuppression by viruses (e.g. HIV) results in increased rates of cancer incidence (Kaposi's sarcoma), HIV does not cause cancer directly. This suggests that immune systems hold these infections in check but profound immunosuppression

• **Figure 9.3** Oncogenesis from integrated virus genes. The effects of proviral genome on gene transcription can occur in at least three ways. (A) The viral promotors cause read through on virus genes including a v-*onc*. (B) A virus without an oncogene may still activate c-oncogenes because of the read through driven by viral promotors. (C) A viral gene product enhances transcription of a cellular gene at a distant site

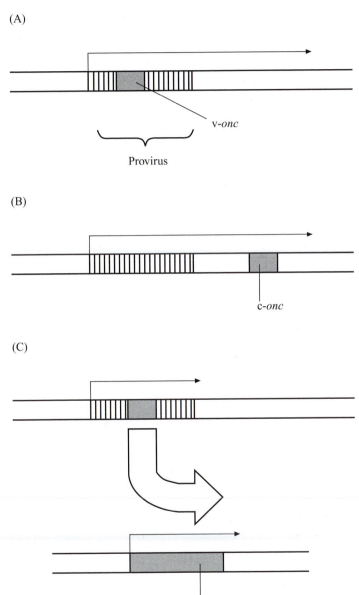

(A)

v-*onc*

Provirus

(B)

c-*onc*

(C)

Cellular gene

cannot. Finally, viruses may act indirectly by increasing replication of cells. Faster-replicating cells are more prone to mutagenic events induced by cofactors, hence HBV may promote replication of liver cells such that other mutagenic agents (e.g. ingested fungal aflatoxins) have a greater effect. Incidentally, a similar sequence of events is postulated to occur in the development of gastric lymphoma following *Helicobacter pylori* infections.

## ■ SUMMARY

Viruses are very successful parasites. Viruses tend to cause asymptomatic infections in humans with only a small fraction of the total population exposed developing symptoms. In patients who become ill, they are usually most infectious just prior to showing symptoms so that the infection is not displayed to the next host and transmission is not compromised. Important differences are seen in DNA and RNA viruses, most notably in acute and persistent infections. Viral infections evoke immune cells to release cytokines in order to curtail the infection and mobilise the immune system. Viruses have numerous mechanisms to modulate the immune response and remain within the host (latency). Such adaptations have resulted in viruses being significant factors in the development of carcinogenesis.

## RECOMMENDED READING

Alcami, A. and Koszinowski, U.H. (2000) Viral mechanisms of immune evasion. *Mol. Med. Today* 6, 365–72.

Collier, J. and Oxford, J. (2000) *Human Virology*, 2nd edition, Oxford University Press, Oxford, UK.

Domingo, E. and Holland, J.J. (1997) RNA virus mutations and fitness for survival. *Ann. Rev. Microbiol.* 51, 151–78.

Everett, H. and McFadden, G. (1999) Apoptosis: an innate immune response to virus infection. *Trends Microbiol.* 7, 160–5.

Kalvakolanu, D.V. (1999) Virus interception of cytokine-regulated pathways. *Trends Microbiol.* 7, 166–71.

Porterfield, J.S. (1992) Pathogenesis of viral infections; in McGee, J.O.D., Isaacson, P.G. and Wright, N.A. (eds) *Oxford Textbook of Pathology*, Oxford University Press, Oxford, UK.

Villarreal, L.P., Defilippis, V.R. and Gottlieb, K.R. (2000) Acute and persistent viral life strategies and their relationship to emerging diseases. *Virology* 272, 1–6.

## REVIEW QUESTIONS

*Question 9.1*  Does the target site of viral replication have any significance on the severity and duration of the infection?

*Question 9.2*  List the reasons to explain why viruses cannot be stated as sole causes of cancer.

*Question 9.3*  What is a viral quasispecies?

*Question 9.4*  Why is the incubation period between 7 and 14 days for many childhood viral infections?

*Question 9.5*  How do RNA and DNA viruses differ in their strategies of human infections?

The impact of fungi on humans is potentially considerable when one considers the possible targets. As parasites fungi primarily infect plants, and the Irish famine of the 1850s in which the potato crop was ruined by *Phytopthera infestans* represents an alarming example of the damaging effects fungal infection has on foodstuffs. Two million people died during the five-year period in which the crops were lost. Less dramatically, fungi also cause spoilage of pharmaceutical and cosmetic products, as well as attacking the timbers of our homes.

Only a small proportion of fungi (<0.1 per cent) exist as parasites of humans and animals, fungi mostly acting as saprophytes (organisms that absorb their organic nutrients from dead substrates). By definition, then, saprophytic organisms are not parasitic because their nutritional sources are not alive; instead the organisms are acting as decomposers and recyclers. As most fungal infections are acquired from an environmental source, fungi are *not* obligate human parasites but, instead, human mycoses can be considered accidental infections that are not designed to facilitate the spread of the organism to new hosts. Hence, human to human spread is very rare for most fungal infections with the exception of the dermatophytes (see below, p. 229).

Clinically, human fungal infections can be grouped into:

(i) **systemic disease** (infections of internal organs of the body), and
(ii) **localised or superficial** infections (infections confined to the skin or mucous membranes that do not invade into deeper tissues or organs – see Table 10.1). This classification is anthropomorphic in that it considers the diseases as they affect man. In order to see what are the important features of the fungi themselves, it is appropriate to group the organisms into diseases by **primary pathogenic fungi** and **secondary pathogenic fungi**. The tiny fraction of fungi that can cause human disease, despite the daily exposure to large numbers of airborne fungal spores, can be taken to mean that most fungi cannot overcome human defence mechanisms.

## ■ 10.1 PRIMARY PATHOGENIC FUNGI

The fungi that are primary pathogens are those that can infect and cause disease in healthy people; 'healthy' in that there are no recognisable predisposing features which

**Table 10.1 Examples of the two major types of human mycoses**

| Mycosis | Aetiological agent(s) |
|---|---|
| **Systemic mycoses** | |
| Histoplasmosis | Histoplasma capsulatum |
| Systemic candidosis | Candida albicans |
| Cryptococcosis | Cryptococcus neoformans |
| Aspergillosis | Aspergillus spp. |
| **Localised infections** | |
| Dermatophytoses | Epidermophyton spp., Trichosporum spp., Microsporum spp. |
| Candidiasis | Candida albicans |
| Pityriasis versicolor | Pityrosporum spp. |

facilitate infection. The diseases that are seen may be systemic infections or superficial infections. This immediately demonstrates how the severity of the infection is not necessarily an indication of the pathogenic potential of the organism.

### ■ 10.1.1 SYSTEMIC INFECTIONS

Fungi that are able to cause systemic illness in healthy people are rare and confined to specific geographic locations across the world. The fungus *Histoplasma capsulatum* causes **histoplasmosis** which is endemic to certain areas in the southern states of America. Another infection is **coccidiomycosis**, caused by *Coccidioides immitis*. Like histoplasmosis, the fungus is found between the southern states of America, Mexico and the northern-most countries of South America. These two infections are caused by inhalation of the fungus, which exhibits **dimorphism**, i.e. can exist as a yeast or a mould. The organisms are acquired by inhalation of the conidia from soil, and develop in the lungs as yeasts. The ability to switch between forms is seen following the change in temperature so that *Histoplasma capsulatum* is a mould when grown at 25°C but grows as yeast at body temperature. This property is then more accurately described as **thermal dimorphism**. The switch from being a mould in the soil where it adopts a saprophytic existence to a yeast when at 37°C presumably favours the parasitic existence in a warm-blooded animal. Dimorphism is a particular feature of the endemic fungi that cause systemic infections but it is not restricted to pathogenic fungi (Table 10.2).

**Table 10.2 Fungal morphology as saprophytes and pathogens**

| Mycosis | Fungus | Saprophytic form | Parasitic form(s) |
|---|---|---|---|
| Coccidiomycosis | Coccidioides immitis | branching hyphae | spherules (sacs of endospores) |
| Histoplasmosis | Histoplasma capsulatum | branching hyphae | yeast |
| Dermatophytosis | Dermatophytes | branching hyphae | branching hyphae |
| Cryptococcosis | Cryptococcus neoformans | yeast | encapsulate yeast |
| Candidiasis | Candida albicans | yeast | yeast, hyphae, pseudohyphae |

## ■ 10.1.2 SUPERFICIAL INFECTIONS: THE DERMATOPHYTOSES

The infections caused by the group of fungi collectively termed the **dermatophytes** are best known by the infection ringworm. Nothing to do with worms, ringworm is a fungal infection of the skin which shows up as a circular area of redness in which the fungal mycelia are radially extending outwards through the superficial skin squames. The sites of infection are reflected in the medical terms tinea barbae (beard), tinea capitis (head), tinea corporis (body), tinea cruris (groin), tinea pedis (foot) and tinea unguinum (nail). The different sites of infection on the body reflect the ability of dermatophytes to invade keratinised tissues (skin, hair and nails) but no further, since these fungi have the ability to obtain nitrogen from the keratin present only in the (dead) tissues in the body.

The dermatophytes are a group of approximately 40 species that are classified by their anamorphic (asexual) stages. They fall into three genera: Trichophyton, Microsporum and Epidermophyton within the Fungi Imperfecta (Deuteromycota). The ability to reproduce sexually is very rarely found in these organisms, and this is thought to represent the intermediate transfer from environmental organisms to a strictly parasitic existence on humans and animals. The organisms can be described according to their ecological location:

* **anthropophilic dermatophytes**: exclusively infect humans,
* **zoophilic dermatophytes**: animal parasites that accidentally infect humans,
* **geophilic dermatophytes**: soil inhabitants that can infect humans and animals.

As zoophilic dermatophyte infections are transmissible from animals to humans, anthropophilic infections are transmissible from person to person, a property unusual in human mycoses (Figure 10.1). In addition, the organisms vary in their geographical

• **Figure 10.1** Lifecycles of dematophytes. Humans are accidental hosts for zoophilic and geophilic dermatophytes but the only recognised host for anthropophilic infections

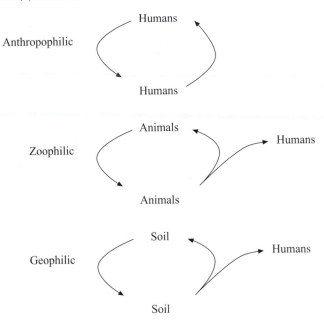

distribution with some widespread (globally) and others restricted to specific areas of the world.

As the dermatophytes infect only keratinised tissues, they do not invade living tissues. In keeping with the descriptions of radial growth in Chapter 4, the lesions radiate as a ring with the advancing hyphae migrating into new tissues, and the pathology shows varying degrees of cell-mediated inflammation with an increased rate of division and keratinisation of the skin layers. This increased cell turnover will provide an increased supply of nutrients for the fungus. The infections with the most degree of inflammation are the zoophilic and geophilic species, unlike the anthopophilic infections, which are often asymptomatic. Such findings support the idea that the most host-adapted species (the anthropophilic species) are better suited to maintaining itself within a human population.

## ■ 10.2 SECONDARY FUNGAL PATHOGENS

The three fungi that we will consider are all able to cause systemic infections, but it appears that predisposing factors in the host are required for secondary pathogenic fungi.

It is because of the advances in medicine that opportunistic infections have become a greater problem, both in incidence and clinical management of the patient. The increase in the number of successful organ transplants, for example, means that such patients are at risk of developing infections due to opportunist organisms, of which fungi have been particularly problematic. We will consider three fungi that have become regular culprits in causing infections in patients with underlying disease: *Cryptococcus neoformans*, *Aspergillus spp.* and *Candida albicans*. In all three organisms, the increased opportunities for invading the patient have arisen from the following factors:

* increased numbers of patients with immunosuppression: naturally occurring such as AIDS or through more successful medical immune manipulation during transplantation.
* continued use of broad spectrum antibiotics.
* increased use of intravenous lines.
* increased rates of survival in premature babies.

### ■ 10.2.1 *CRYPTOCOCCUS NEOFORMANS*

*Cryptococcus neoformans* is a yeast that has become a frequent invader of patients with impaired cell-mediated immunity, whereas clinical disease is rare in healthy people. In such patients the yeast often invades the central nervous system and is particularly frequent in causing meningitis, although pneumonia is not uncommon. The organism can be recovered from soil throughout the world, thus differing from the primary pathogens described above which have a restricted geographical location. Cryptococci will favour nitrogen-rich soil, typically achieved through bird droppings. There are approximately twenty species of cryptococci but only *C. neoformans* causes disease in humans.

Unusual amongst fungal pathogens is the production of a polysaccharide capsule by *C. neoformans*. All cryptococci can produce polysaccharide capsules but, when growing in soil, the capsule is usually not evident. However, when nitrogen or moisture levels drop, capsule synthesis is stimulated. In this way the capsule can provide a reservoir of water during an oncoming period of drought and protect the yeast from dehydration. *C. neoformans* will grow at 37°C but most species of cryptococcus do not tolerate this temperature, preferring ambient temperatures. Birds typically have body temperatures over 40°C and so, whilst cryptococci might be carried in the avian gut, they do not cause illness in birds.

Humans are exposed to the organism via the respiratory tract. The dehydrated yeast cell will be of sufficiently small diameter to act as an aerosol such that the yeasts may enter the respiratory tract, whereas yeast cells with abundant capsule present will be too large to be effective aerosols. The increase in carbon dioxide and fall in nitrogen concentrations that the yeast encounters once inhaled stimulate the synthesis of the capsule. These changes are good examples of the environmental signalling discussed in Chapter 8. *Cryptococcus neoformans* will multiply within phagocytes, resistant to the antimicrobial attack of the phagolysosome. As with bacterial capsules, the crytococcal capsule has a number of functions: the polysaccharide capsule is a polymer of **mannan**, a typical component of fungal cell walls, which is poorly immunogenic. The mannan appears to inhibit effective phagocytosis by the host neutrophils unless specific antibodies are produced. The capsule is also a weak activator of the host immune response if not intrinsically immunosuppressive. Shed capsular material will also act to bind host antibody and complement components, thereby protecting the organisms itself.

Additional virulence factors have been suggested and the production of a phenoloxidase enzyme may contribute to the neurotropism of the yeast. The phenoloxidase enzyme catalyses the conversion of neurotransmitters such as dopamine, which are particularly abundant in brain tissue, into melanin. Melanin acts as a scavenger for reactive oxygen species such as superoxide ion. This may protect the yeast from such damage due to reactive oxygen species by leucocytes.

To keep perspective, it should be remembered that *C. neoformans* is rarely seen as a cause of disease in healthy people despite the fact that many people are likely to encounter the organism. Only patients with impaired or defective immune systems are likely to become ill with this organism. Interestingly, patients with deficiencies in their cell-mediated immunity develop infections with this organism. These patients are typically people with late stage AIDS and patients with immunosuppressive treatment for organ transplant or autoimmune disease like Hodgkin's disease. The role of the humoral arm of the immune system is less clear. Significant antibody titres in serum do not appear to provide protection in these patients, which supports the critical role of the cell-mediated arm.

## ■ 10.2.2 ASPERGILLUS INFECTIONS

Aspergilli are moulds and, consistent with a free-living fungus, they are ubiquitous in the environment where they can be cultured from soil, compost and grain. They are also easily cultured from foods and plants in indoor environments. The abundance and small size of the spores produced (2–3 μm) means that most people will encounter them as aerosols, able to penetrate the lower respiratory tract. Fortunately, healthy people can deal with the spores, i.e. destroy or eliminate them, and hence do not become ill, but the groups of patients who develop systemic illness with opportunistic fungi are sufficiently immunocompromised that the spores are able to establish themselves within the host and germinate into hyphae. Typically, Aspergillus infections occur in patients with prolonged and severe neutropenia (very low numbers of neutrophils in the blood) and often colonise pre-existing lesions such as old tubercular lesions or lung cancer sites. Of the numerous species within the genus, *Aspergillus fumigatus* is the most frequent species isolated from patients with disease. Quite what distinguishes this species from the rest, in terms of virulence capacity, is not known.

### ■ 10.3.2 ANATOMICAL STRUCTURES

The ability of certain fungi to use specialised structures to penetrate the host tissues is widespread amongst plant pathogens and also occurs in human fungal pathogens. The ability to switch between hyphal and yeast forms (dimorphism) is commonly observed when infecting animal hosts in the course of an infection. Candida develop hyphae (pseudohyphae) when invading human tissue. In contrast to bacterial pathogens, the use of capsules as virulence factors is almost unknown in fungal pathogens with the exception of *Cryptococcus neoformans*.

### ■ 10.3.3 SECRETION OF NUMEROUS HYDROLYTIC ENZYMES

The heterotrophic nature of fungal nutrition means that they secrete numerous degradative enzymes in large quantities, many of which may attack host tissues. Consequently, it is difficult to assess experimentally the contribution of individual enzymes as the other enzymes may compensate for any loss. Furthermore, the property is characteristic of most fungi, yet very few are pathogenic to humans.

### ■ 10.3.4 EVADE IMMUNE RESPONSES

Fungal pathogens that persist in human hosts do so by various mechanisms:

1. infection of immuno-priviliged sites (skin, central nervous system),
2. fungal spores are relatively resistant to phagocytosis compared with hyphae,
3. release of immuno-modulatory substances or molecular mimics.

This chapter has outlined four fungal infections to illustrate the variety in conditions that underlie these events. They highlight the need to consider both the state of the host and the fungus in conjunction; viewing either in isolation is insufficient. With fungal infections of humans that are accidental to the normal lifecycle of the organism, the virulence factors are unlikely to be specific to that organism but instead common to most fungi. The reason disease has developed is because of an accidental encounter. Accidental human mycoses have little opportunity for transmission to new hosts in sharp contrast to well-adapted pathogens, the dermatophytes and *Candida albicans*. Studies on the pathogenesis of fungi have lagged behind those of bacteria and viruses, although the rising incidence of mycoses will stimulate much needed funds for new

---

### ■ BOX 10.1 TOXIC PRODUCTS

Whilst we are primarily concerned with how microbes infect humans, it is important to remember that certain human diseases result from fungal growth in products that we consume. Fungal poisoning of foodstuffs can be separated into:

- **mycetism**, the ingestion of poisonous mushrooms, and
- **mycotoxicosis**, the ingestion of food products containing toxins (**mycotoxins**) secreted by fungi that have contaminated the raw material or product. The mycotoxicoses include **ergotism**, in which cereal grain is infected with *Claviceps purpurea* (and was once a common illness known as 'St Anthony's fire') and ingestion of **aflatoxin**. Aflatoxins are potent carcinogens produced by *Aspergillus spp.*, linked with the development of liver cancer. Peanuts are often found to contain high concentrations of these toxins as a result of fungal contamination during storage.

Finally, we should perhaps include alcoholism as a widespread type of mycotoxicosis!

studies. However, the methodology will be more complicated than with bacteria as genetic complexities of most fungi will complicate the 'virulence gene' deletion studies.

## ■ SUMMARY

As saprophytes, fungi are mostly accidental pathogens of humans. Primary fungal pathogens (certain dimorphic fungi and the dermatophytes) are able to infect humans without recognisable predisposing factors in the host. Primary pathogens cause systemic or superficial (skin) infections. The dimorphic fungi responsible for systemic infections are acquired via inhalation of spores and are able to replicate in the niche (human tissue). The superficial dermatophyte infections are restricted to keratinised tissues. Secondary fungal pathogens replicate in humans only through predisposing factors such as immunosuppression or traumatic implantation of the organisms into the tissues but, as a consequence, are more likely to cause systemic disease. The virulence factors recognised include dimorphism, thermal tolerance and immunomodulatory abilities, but increasing numbers of immunosuppressed hosts through medical improvements or disease such as AIDS means that fungal infections are increasing in number. This increase indicates the importance of host condition in the balance between fungal and human factors in whether infections are established.

## RECOMMENDED READING

Cutler, J.E. (1991) Putative virulence factors of *Candida albicans*. *Annu. Rev. Microbiol.* 45, 187–218.

Denning, D.W. (1991) Epidemiology and pathogenesis of systemic fungal infections in the immunocompromised host. *J. Antimicrob. Chemother.* 28, Suppl. B, 1–16.

Fridkin, S.K. and Jarvis, W.R. (1996) Epidemiology of nosocomial fungal infections. *Clin. Microbiol. Rev.* 9, 499–511.

Hogan, L.H., Klein, B.S. and Levitz, S.M. (1996) Virulence factors of medically important fungi. *Clin. Microbiol. Rev.* 9, 469–88.

Kobayashi, G.S. and Medoff, G. (1998) Introduction to the fungi and mycoses, in Schlaecter, M., Engelberg, N.C., Eisenstein, B.I. and Medoff, G. (eds) *Mechanisms of Microbial Disease*, 3rd edition, Lippincott, Williams & Wilkins, Baltimore, USA.

Richardson, M.D. (1992) Fungal infections, in McGee, J.O.D., Isaacson, P.G. and Wright, N.A. (eds) *Oxford Textbook of Pathology*, Oxford University Press, Oxford, UK.

Weitzman, I. and Summerbell, R.C. (1995) The dermatophytes. *Clin. Microbiol. Rev.* 8, 240–59.

## REVIEW QUESTIONS

*Question 10.1*     Which are the two most at risk population groups for fungal infections?

*Question 10.2*     To what extent is *Candida albicans* representative of fungi?

*Question 10.3*     Many dermatophytes are anamorphic. Explain the significance of this sentence.

*Question 10.4*     What is dimorphism?

# CONTROL OF MICROBIAL INFECTIONS

This chapter considers some of the options available to help to control the impact of infectious diseases on populations. The study of diseases within populations is called epidemiology. For microbiologists, infectious disease epidemiology is an increasingly important area of study as infectious diseases do not appear to be diminishing in number. The increased mobility of people across the globe has meant that infectious diseases can be acquired in one part of the world and within hours transported to another. The movement of micro-organisms will be the same. To keep an eye on the changes in distribution and numbers of infectious diseases, national centres are required. Within the USA this is based at the Center for Disease Control (CDC), Atlanta and in the UK, the Central Public Health Laboratory, Colindale, London. This overview of national trends is part of the surveillance of infectious disease. The reason that certain infectious diseases must be notified to the public health department is because of the value of national surveillance in identifying and controlling epidemics. Box.11.1 gives the important terms that are used.

## ■ 11.1 EPIDEMIOLOGY OF INFECTIOUS DISEASE

What are the features of micro-organisms that spread as infectious diseases through a community? Classic infectious diseases such as measles can be modelled assuming that

---

**■ BOX 11.1 EPIDEMIOLOGICAL TERMS**

**Incidence**: number of new cases of infection occurring within a population over a defined period of time ('per unit population per unit time').

**Prevalence**: the total number of cases in a defined population either at any one point in time (point prevalence) or over a defined period (period prevalence). Prevalence gives an indication of how widespread the infection is within a population, whereas incidence, by recording only the new cases, tells you how rapidly the outbreak is growing.

**Outbreak**: the incidence of an infection increases within a population above that normally expected. Only two cases represent an outbreak when considered against a background of an infection that is normally an uncommon occurrence. An outbreak is thus defined relative to the normal incidence. A **sporadic case** is a case of illness (infection) not considered part of an outbreak.

the infection is transmitted randomly between infectious and susceptible people. The rate of transmission will be proportional to the product of the number of susceptibles, multiplied by the number of infectious people and is based on the mass action principle (Figure 11.1). If $S$ is the number of susceptibles and $I$ is the number of infectious people, then the rate of new infections ($N$) will be the product of the rate of mixing and the concentration of the two components (susceptibles and infected):

$$N = \beta SI$$

Mathematically, the proportional rate of increase means that there must be a constant ($\beta$) that moderates the product. That constant represents the probability of effective transmission. For example, the efficiency of transmission of hepatitis B virus being transmitted by sneezing is extremely low in comparison to measles virus, which has a high value of $\beta$ (closer to one). The transmission coefficient ($\beta$) will vary between micro-organisms and their route of transmission. The other two components $I$ and $S$ are more direct: the more infectious people ($I$) or susceptibles ($S$) there are, the greater the rate of new infections occurring ($N$).

• **Figure 11.1** The law of mass action. Originally used to describe the motion of two gases in a closed container, the law of mass action can be used to help model ideas about the mixing of an infectious agent such as measles between susceptible and infected people. (A) shows the two populations, susceptible (S, open circles) and infected people (I, shaded circles) moving around at a fixed rate (arrows). The rate of new infections (N) will be the product of S and I and the transmission coefficient (β, see text). The rate of new infections can be increased by either (B) increasing the number infected, or susceptible, as well as (C) increasing β, the efficiency of transfer of the organism (indicated by increased length of the arrows)

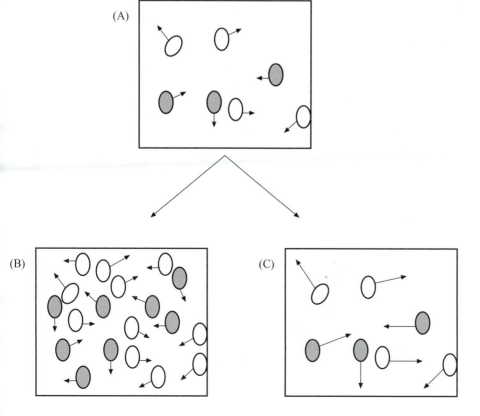

The number of people directly infected by someone who is infected by a particular micro-organism in an entirely susceptible population is termed $R_o$, the **basic reproductive rate**. Assuming that the infected and susceptibles can mix randomly (**homogenous mixing**) we can use the mass action principle to define $R_o$ as the product of the susceptibles ($S$), the probability of effective transmission ($\beta$) and the period of time each case is infectious ($D$).

$$R_o = S\beta D$$

The equation again shows that increasing any component will increase $R_o$, e.g. increase the duration of infectiousness and the number of people who become infected per case increases. The conditions defined here give a numerical value to the incidence of new cases in a population consisting only of susceptible and infected people, who mix randomly. These conditions are somewhat artificial because the proportion of people who are immune is not considered, but nevertheless, such values help in analysing the behaviour of different micro-organisms. Highly infectious diseases like measles have high values of $R_o$ (around 20) whereas poorly infectious diseases (famously leprosy) have $R_o$ values much closer to 1. Being highly infectious, organisms with high $R_o$ infect people at a younger age than organisms with higher $R_o$. In other words, $R_o$ and average age at infection are inversely related.

For the organism, three key limits are predicted:

$R_o < 1$: the infection declines and disappears from the community.
$R_o \cong 1$: the infection persists, i.e. remains endemic.
$R_o > 1$: the infection will increase in incidence and spread as an epidemic.

When an infectious disease with a high $R_o$ such as measles invades a completely susceptible population (an isolated island community) around 20 new cases arise from every infectious person and a short-lived epidemic will occur. Fortunately it will burn itself out as the number of susceptibles is quickly exhausted. The epidemic will occur more rapidly than with an organism of lower $R_o$ since, once infected, people will become immune (or perhaps die), the number of susceptibles falls and the virus finds fewer and fewer people to infect. This phenomenon is critical in terms of the outbreak. When $R_o < 1$, less than one secondary case is produced for each infected individual and the epidemic fades. Thus the **actual** or **effective reproductive rate** ($R$) falls during the course of an outbreak because the number of susceptibles that become immune is not replenished. The effective reproductive rate is therefore variable and related to the basic reproductive rate thus:

$$R = R_o x$$

where $x$ is the proportion of the population susceptible, e.g. if $R_o$ is 20 and 50 per cent of the population are immune, then $R = 20 \times 0.5 = 10$.

Knowing that $R_o = 1$ represents a threshold point below which the epidemic will die out, we have obtained a theoretical minimum number of susceptibles required to maintain the organism within the community. From our equation, this threshold – the **numerical density threshold ($D_T$)** – can be predicted by setting $R_o$ to 1.

$$1 = S\beta D$$

Therefore:

$$1/S\beta = D_T$$

$D_T$ will be high for organisms with high $R_o$ because they spread so effectively through the population and therefore need high densities of susceptibles.

The outcome of our proposed measles epidemic can therefore be predicted to some extent. The key parameters in an epidemic are:

- the size of the population, and,
- the rate of new susceptibles appearing.

The epidemic will fade and disappear as $R$ falls to less than unity, but if the population density is high enough ($>D_T$) the organism may persist and reappear later as a second epidemic. This pattern may repeat and give epidemics every couple of years, following the oscillation of $R$ around unity (Figure 11.2). The length of time between epidemics will depend on the arrival of new susceptibles (such as births or migrations).

Should all organisms seek a high $R_o$? Not necessarily. Microbes with high $R_o$ run the risk of quickly eliminating all potential hosts as they sweep through the community. They may be viewed as too effective, but in large, dense cities this strategy is workable. More patient strategies are needed if the population is sparse. Production of robust and

• **Figure 11.2** Effect of introducing vaccination on epidemics. The effect of the introduction of a vaccination campaign with coverage levels below that needed to eradicate the disease. (A) shows the inter-epidemic period lengthens but the disease does not disappear altogether because the number of susceptibles has not been reduced sufficiently to eradicate the disease completely. (B) shows how the age range changes upon introduction of a vaccination campaign. Note the increase in age of infection in the vaccinated population (dotted line) compared with the unvaccinated population (entire line)

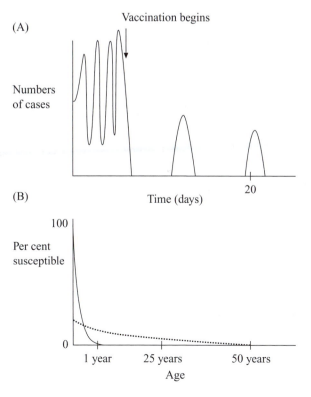

long-living infective forms is one option. Bacterial endospores are a good example. Viruses are less well equipped to cope with long periods outside a host and some have adopted vertical transmission as the ultimate adaptation to ensuring the transmission of the organism when low-density populations exist (e.g. chronic Hepatitis B infection).

### ■ 11.1.1 THE VALIDITY OF $R_0$

One of the values of $R_0$ lies in what its composition tells us. The characteristics that comprise $R_0$ must be essential and important for the organism, therefore reduce any of these values and the efficiency of spread will be impaired. If any of the parameters are reduced to zero, in theory the epidemic will be halted.

If $R_0 = S\beta D$ then the number of susceptibles ($S$), the efficiency of transmission ($\beta$) and infectious period ($D$) will all be suitable targets for control of the epidemic. The easiest target will often be to reduce the number of susceptibles through vaccination.

As a marker of the ability of an organism to spread, $R_0$ gives an indication of the propensity of an organism to invade a population (cause an epidemic). Organisms of high $R_0$ will spread very efficiently and infect the very young. $R_0$ can be used to predict the levels of vaccination needed to eradicate infectious disease. The effective reproductive rate $R$ must be less than one for the disease to disappear. Earlier, we defined $R = R_0 \cdot x$, where $x$ is the fraction of the population susceptible. If a completely susceptible population is represented as unity, then the actual fraction of the population immunised can be represented as $(1 - p)$, hence:

$$R = R_0 (1 - p)$$

Substituting $R$ with 1 yields:

$$1 = R_0 (1 - p)$$

This can be rearranged to find the percentage ($p$) that need to be vaccinated as follows:

$$1/R_0 = 1 - p$$

$$1/R_0 + p = 1$$

$$p = 1 - 1/R_0$$

$p$ will be the boundary condition, i.e. it sets the minimum proportion of the population that needs to be vaccinated in order for the effective reproduction rate to fall to one. It follows that vaccination rates must not be less than that set by $1 - 1/R_0$.

To combat measles it is necessary to achieve very high vaccination rates (around 95 per cent uptake) in children as young as possible. It is clearly not that helpful to vaccinate children at an age that is older than the average age of infection.

Does the incidence of infection drop with increasing vaccination rates? Mathematical modelling indicates that the relationship is not linear, as shown in Figure 11.3. Only once $R$ has been reduced to values approaching one do noticeable reductions in infection result. Conversely, however, the value of $p$ required to eliminate an infectious disease does not have to be 100 per cent coverage. The protection of the few unvaccinated by the immunity of the bulk of the population is called 'herd immunity'. The protection lies in the absolute numbers of susceptibles falling below the critical threshold value set by $D_T$.

**• Figure 11.3** The non-linear relationship predicted between the proportion of the population vaccinated and the reproductive rate of an infectious disease. The slope of the graph shows how vaccination will reduce the incidence of the disease with greatest effect when the effective reproductive rate approaches 1. Note also that the uptake of vaccination needs to be greater than 95 per cent to see an infectious disease with a high $R_0$ (such as measles, dashed line) eliminated but less with a disease such as poliomyelitis with a lower $R_0$ (dotted line)

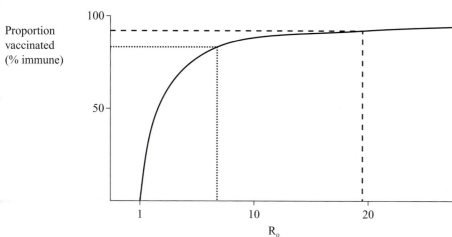

What measures can be implemented to reduce β and D? For highly infectious diseases such as measles the idea of mass action as the basis for describing the random mixing of infectious and susceptible people appears to be a good model for describing what actually happens. Hence, $R_0$ proves useful in comparing the infectiousness of different organisms. With measles, diphtheria and rubella, each confer lifelong immunity. This characteristic helps in monitoring a population for immunity as the protective antibodies, all developed against a single dominant antigenic type, can readily be measured in the serum of people. The variation in antigenic types seen with influenza virus make predictions of the virus almost impossible, as well as the manufacture of correct vaccine that will be protective every year. $R_0$ can be estimated from serological surveys of the population.

The age at which antibodies appear can be used to calculate $R_0$ (the younger the age at which the infection occurs equates to a higher $R_0$ and vice versa). Again, such methods are problematic for influenza virus because different strains of influenza virus may circulate through the population periodically, making the serological testing complicated.

Not all infectious diseases spread horizontally like measles. Hepatitis B virus is acquired parenterally or vertically. Such transmission is not random; the chances of all individuals in the population encountering each other are not equal, thus mass-action models do not apply. An additional difficulty is that the efficiency of transmitting the infection may also vary between different groups of people. Sexually transmitted diseases are an example of a non-homogenous mixing of the population (celibate people will not become infected) as well as examples of differences in the efficiency of transmission (transmission of HIV and Hepatitis B virus is less efficient in heterosexual partners than in homosexuals).

■ **11.2 VACCINATION**

What measures exist to prevent and reduce the incidence of infection? Arguably the provision of clean drinking water, nutrition, sanitation and general economic health have a profound effect, but to these vaccination must be added. Only vaccination has resulted in an infectious disease being eradicated from the human population, with the global vaccination against smallpox declared successful by the World Health Organisation in 1979.

The mechanisms by which vaccination works in humans is essentially immunology and will not be covered here. The area of interest to us is what vaccination tells us about infectious diseases of humans. The success of vaccination depends foremost on our understanding the physiology and ecology of the organism in question.

Whilst vaccination protects the individual, there are benefits to the whole community when vaccination uptake rates are high. The few remaining individuals that have not been vaccinated are protected by the **herd immunity** principle. As discussed previously, $D_T$ represents the minimal number of susceptibles that are needed for a particular organism. Herd immunity arises if the number of susceptibles fall below that value.

Herd immunity in humans only applies to infectious diseases that are strict human pathogens. Any infection that is acquired from the environment or from animals will be unaffected by the reduction in susceptibles ($D_T$) because it has a natural reservoir distinct and separate from humans (i.e. the environment or animals). For example, the reservoir for tetanus bacilli is the soil and rabies virus is a zoonosis. Neither are caught from other humans. Thus, vaccination of other people provides no protection to the individual who has caught tetanus from a rose bush thorn or rabies from an infected dog. Note how influenza virus epidemics cloud the distinction. Influenza originates from an avian reservoir (notably ducks) and once transferred to humans, spreads as epidemic influenza from person to person. Strictly, this qualifies influenza as a zoonosis. The vaccination coverage is selective, i.e. targeted to the most susceptible, such as elderly people with chest disease. In this circumstance vaccination rates are unlikely to ever impact on the $D_T$, hence effective reproductive rate ($R$) is unaffected and herd immunity is unlikely to occur.

Ideally, a vaccine should be **stable** enough to withstand being transported across the country without loss of potency. One of the reasons for the success of the smallpox vaccine was its stability at high temperatures whilst being delivered to the populations close to the Equator. Also, a vaccine should be totally safe and effective, cheap and easy to administer (preferably oral) and provide lifelong immunity. The reality is that some vaccines have many of these features but none possess them all. Despite this vaccines have reduced the incidence of infectious disease, reduced the incidence of lifelong sequelae and disability and reduced the mortality rate.

DDD, the value of vaccines:

- Reduce Disease,
- Reduce Disability,
- Reduce Death.

The types of vaccine that are produced commercially can be separated according to whether the vaccine uses intact organisms, either live or killed, or components/products of organisms. What follows are selected examples and important features of the different types of vaccine currently used.

### ■ 11.2.1 HETEROLOGOUS VACCINE

The use of cow pox virus as a vaccine for smallpox by Edward Jenner is an example of a live, heterologous vaccine. The term 'heterologous' is used to describe how a microorganism (cow pox caused by *Vaccinia* virus) that causes infections in other hosts (cows) is able to induce protective immunity to the human form of the disease (smallpox caused by *Variola* virus) if used as a vaccine.

### ■ 11.2.2 LIVE, ATTENUATED VACCINE

The BCG is a live strain of *Mycobacterium tuberculosis* that has been cultivated on culture medium containing bile, thereby reducing the virulence of the organism compared with the wild type. The Sabin polio vaccine is another example of a live, attenuated vaccine.

### ■ 11.2.3 KILLED ORGANISMS

Influenza and rabies vaccines are killed organisms, achieved by treating the virus with β-propiolactone, a chemical not dissimilar to formaldehyde, which does not destroy the protective antigens on the virus. The difference between influenza and rabies vaccine is that rabies is a monotypic virus (has one antigenic type), unlike influenza virus which undergoes antigenic modulation and requires a new vaccine for each type.

### ■ 11.2.4 SUBUNIT VACCINES

Inactivated intact bacteria have been largely replaced (where possible) by the use of the appropriate component that provides protective immunity. This reduces the risk of any contamination of vaccine by organisms that are not killed. The protective antigens on an organism are likely to be exposed on the surface. Several vaccines use the appropriate structures in place of intact organisms, e.g. the polysaccharide capsules from *Haemophilus influenzae*, *Neisseria meningitidis* and *Streptococcus pneumoniae*.

### ■ 11.2.5 TOXOIDS

Toxoids are the secreted exotoxins from toxigenic pathogens such as *Clostridium tetani* (tetanus toxin) inactivated by formaldehyde. The formaldehyde treatment inactivates the toxin but does *not* destroy the antigenic determinants that evoke the immune response. In this case the antigens that protect against the disease are coded on the toxin rather than a surface structure of the organism. Such vaccines prevent clinical illness developing but not necessarily infection.

### ■ 11.2.6 PASSIVE VACCINES

Passive vaccines differ fundamentally from the other types of vaccine. Passive vaccines are simply injections of pre-formed immunoglobulin raised against the organism in question. Hepatitis A virus is usually acquired from infected seafood, particularly mussels and other filter feeders. The virus is passed into the sea water from faeces of infected patients but are concentrated in the mussels growing in the contaminated water. People travelling to areas of particular high risk of Hepatitis A infection are given anti-hepatitis A immunoglobulin as immunoprophylaxis. Such a procedure provides protection for the individual only, driven by the particular circumstances rather than national campaigns. The protection is finite, lasting only the duration of the immunoglobulin in the bloodstream (approximately 3 months).

### ■ 11.2.7 WHAT DO THESE VACCINES HOPE TO ACHIEVE?

With smallpox vaccine, global eradication was achieved following a WHO campaign that chased all cases of the disease across the world. Usually vaccination is used to limit the impact (morbidity and mortality) of an infectious disease rather than eliminate it. It is only realistic to attempt eradication of an infectious disease that has no inanimate or animal reservoir, i.e. a strict human pathogen. Even many strict human infectious diseases may be impossible to eradicate because of antigenic variation in the organism, or short-lived protection against the disease. For infectious diseases that cause childhood epidemics, such as measles, vaccination seeks to protect the community as well as the individual. Vaccination is also used for individuals at high risk of acquiring specific infections, usually related to their work. For example, vaccination against anthrax for people who work with animal skins/hides/carcasses or vaccination against hepatitis B virus for healthcare workers. Such specific vaccination has no expectation of reducing the incidence of the disease in the population at large, nor of providing herd immunity. Instead, these people are vaccinated because they are at greater risk of encountering the organism.

Vaccination against certain infectious agents may be futile. It is suggested that chickenpox (*Varicella* virus) is usually sufficiently mild when caught as a child and vaccination simply will be an unnecessary financial cost and the associated vaccine-related side effects will increase.

### ■ 11.2.8 THE PROBLEMS OF VACCINATION

There is no such thing as a perfectly safe medical procedure. Ironically, vaccination suffers from poor public perception as a result of its effectiveness in eliminating the infectious disease. By eliminating whooping cough and all the associated crippling illness that *Bordetella pertussis* creates, the only events that attract attention are adverse reactions in a small proportion of children vaccinated. Vaccine-induced disease will always accumulate in absolute numbers against a background of falling or absent numbers of cases.

If the vaccination uptake rate has not reached the threshold level for $R = 1$ then the organisms will still circulate (i.e. remain endemic) in the population. In these circumstances vaccination can be envisaged to increase the average age at which people become infected because it is taking longer for the organism to encounter the susceptible fraction of the population (Figure 11.2b). With rubella virus (German measles) infections in older children may coincide with pregnancy. Pregnant women that contract rubella virus in the first three months of pregnancy are at risk of the virus also infecting the developing foetus. Such an infection causes permanent damage such as impaired vision, deafness, cardiac defects and other abnormalities (congenital rubella syndrome, CRS). Thus, vaccination rates below that required for eradication will inadvertently place women at increased risk of CRS. A less catastrophic consequence is seen with mumps virus infections in men. With an increased age at infection, the illness has more serious effects such as orchitis (inflammation of the testicles, with possible temporary sterility) and even meningitis.

Maintaining the necessary high vaccination rates against infectious diseases such as measles requires continued public health momentum. Complacency will occur if the disease is not seen for several years and people will forget or relegate the childhood disease to history. Compulsory vaccination in order to attend school has been one approach to counter this problem.

Vaccines need to cover the window of susceptibility in newborn children. The maternal antibodies provide immunity for the first 3 months but the levels of circulating immunoglobulins in the newborn child will disappear over time. The period that follows (3–12 months of age) is when the child is most vulnerable to acquiring highly infectious diseases such as measles and *Haemophilus influenzae* capsule type b. Unfortunately, vaccines tend to be less efficient in children of this age range (for a variety of reasons such as poor immunogenicity of the vaccine target and residual maternal antibodies). This leaves the child vulnerable. The concept of timing of vaccination obviously means that vaccines must be given before the average age of infection.

Microbes that undergo antigenic variation will be under selective pressure such that the antigenic variant that is not in the vaccine will be selected for. The continual surveillance of influenza virus for new antigenic types is the only option for creating an appropriate vaccine until a generic cross-protecting antigen can be found.

**Table 11.1 If vaccination against smallpox worked, why not measles?**

|  | Smallpox | Measles | Poliomyelitis |
|---|---|---|---|
| Reservoir: | Humans | Humans | Humans |
| Acute infection (no persistence) | Yes | Yes | Yes |
| Number of serotypes | 1 | 1 | 3 |
| Infectious during prodrome | No | Yes | Yes |
| Subclinical cases | No | Yes | Yes |
| $R_o$ | <5 | 12–20 | 5 |
| Average age of infection | 15 yrs | 2–5 yrs | 10 yrs |
| Percentage vaccination uptake for elimination ($p$) | 80% | >90% | >80% |

Table 11.1 indicates why smallpox was such a good target for eradication. A stable, one-course vaccine could be given well before the average age of infection. The smallpox virus has a low $R_o$ and, accordingly, a high average age at first exposure to the virus. The low $R_o$ results in fewer people needing to be vaccinated to eliminate the organism. The virus has no animal reservoir, is antigenically stable and a short infectious period results in a readily identifiable illness (the horrid lesions covering the entire face of smallpox victims is unmistakable). Thus, the disease is easy to trace, a factor that helped during the global vaccination campaign. Measles will be more difficult to eradicate but has many of the important features that helped eliminate smallpox. Likewise with poliomyelitis.

## ■ 11.3 MICROBIAL CONTROL (OTHER THAN VACCINATION)

The old adage that prevention is better than cure is undeniable because it will be impossible to eliminate pathogenic micro-organisms from our environment (despite the apparent success in eliminating smallpox virus). Rather than attempt to kill organisms in their natural habitat, which will include soil and animal sources, the control of infectious diseases aims to reduce contact between the organism and potential hosts. Three targets can be proposed:

- reducing the source of the infection,
- reducing the opportunities for transmission,
- increasing the resistance of the host.

The measures that a country takes (for it must be a national programme if there is to be any real chance of being effective) are usually the following:

- general public health: provision of clean drinking water and disposal of sewerage,
- control of animals that might cause zoonoses,
- vaccination,
- immuno- or chemo-prophylaxis,
- vector control.

## ■ 11.3.1 GENERAL PUBLIC HEALTH

In the nineteenth century, considerable improvements in public health resulted from the introduction of clean water supplies and public sanitation systems. These, and other improvements that followed (nutrition, housing, education, childcare improvements) drastically cut morbidity and mortality rates in city and rural populations across Europe. Even before the introduction of antibiotics, the infectious disease burden had started to fall in England, a testament to the importance of these measures. With increasing biomedical sophistication it is easy to lose sight of the value of the essential public health measures in favour of costly new innovations. Faeco-oral transmission of *Vibrio cholerae* is prevented when clean water supplies are maintained. Vaccination should not seek to replace or obviate the need for water supplies free of sewerage. The high childhood mortality rates in countries with inadequate water supplies are due to infectious diseases and illustrate the impact of failing in these basic needs. Although entirely preventable by public sanitation, high childhood mortality tends to lead to high population growth and poverty which compound the problems. A word of warning, however, about cleaning up water supplies: poliomyelitis (see Box 11.2)

### ■ 11.3.2 CONTROL OF ZOONOSES

Zoonotic sources of infections will be impossible to eradicate without killing all the animal reservoirs. Alternative strategies are the only hope and, with over 200 organisms that can cause zoonoses in humans, control rather than eradication is of great importance. The transmission of many zoonoses from animals to humans occurs via insect **vectors**, for example louse-borne infections by *Rickettsia spp.* (typhus), tick-borne infections by *Borrelia spp.* (e.g. Lyme's disease), yellow fever virus is transmitted by mosquitoes of the genus *Aedes*, and plague (*Yersinia pestis*) transmitted by fleas. Perhaps the most famous zoonosis, rabies is caught directly from the bite of infected animals, typically dogs. With deforestation and increased expansion of cities (the slum regions, that is!) humans are encountering arthropod vector-borne diseases more frequently. The displaced vectors are driven from forests into new geographical locations and the micro-organisms spread into new vectors.

Less shocking than exotic infections such as Ebola virus outbreaks, food acts as the source and vehicle for transmission of infections such as salmonellosis and campylobacteriosis and thus may be zoonoses. Food products are regulated at several levels but simple hygienic precautions will prevent many infections. Correct storage temperatures will prevent multiplication of microbes and adequate cooking temperatures will

---

### ■ BOX 11.2 POLIOMYELITIS

Whilst the improvements in dealing with urban sewage disposal eliminated faeco-oral diseases such as typhoid and cholera from England, the reduction in exposure to other faeco-orally-spread microbes had an unexpected consequence: an increase in paralytic poliomyelitis. Poliomyelitis virus is an enterovirus; it is excreted from the intestine and transmitted faeco-orally. In developing countries, the virus is widespread and can be readily isolated from sewage. Like other human virus infections the organism is mostly asymptomatic (95 per cent of infections are without symptoms) but the incidence of symptomatic illness, notably paralytic poliomyelitis, increases the older the age of the patient on first exposure. If the virus is circulating widely in the community then the age of exposure is very young. Improvements in sanitation will result in a reduction in prevalence of the virus and the average age of infection increases and, consequently, the incidence of paralytic poliomyelitis increases disproportionately. Thus in contrast to many infectious diseases, improvements in sanitation result in an increase of clinical disease, although the overall incidence of the infection is reduced overall.

kill organisms. Pasteurisation of milk is an outstanding example of the efficacy of disinfection eliminating a disease; sterilisation of milk at extreme temperature levels is not necessary to kill *Mycobacterium bovis*. The control of food-borne disease will depend on removal of the pathogens in the original foodstuff (e.g. salmonellae in chickens and their feed) as well as maintaining correct handling and cooking methods. Each of the various steps in the food chain (source, manufacture, distribution and destination) should offer control points at which microbial contamination can be checked.

### ■ 11.3.3 CONTROL OF VECTORS

The control of vector-borne disease has significantly greater problems than in directly acquired zoonoses such as food poisoning. With the increases in the distribution and number of cases of Dengue fever, an arthropod-borne illness caused by a flavivirus, we have problems similar to those in controlling malaria, notably controlling the mosquitoes that carry the virus. The mosquitoes that carry dengue virus, *Aedes spp.*, have spread back into the central Americas having once been eradicated. This means that Dengue is now the most common arbovirus infection in humans responsible for millions of cases annually. It is clear that eradication is likely to be impossible for any zoonosis because it will be unlikely that the reservoir(s) or vectors can be eradicated, especially if the organism is able to reside in a variety of host and vector species. With dengue virus, the virus is maintained in monkeys and transferred by the mosquito within forests (hence termed **sylvatic cycle**). Disturbingly, the virus has widened its vector range by using the *Aedes* species that prefer urban environments (*Aedes aegypti*) such that dengue is maintained within human populations (the **urban cycle**). Note then that urban dengue is not a zoonosis).

You should note that with malaria in humans, the species responsible (protozoa of the genus Plasmodium) are restricted to, and transmitted only between, humans. Malaria is therefore *not* a zoonosis. Theoretically at least it is possible to eradicate malaria if you eradicate the vector. With zoonoses, eradication of the natural host is likely to be impossible. Under such circumstances one can only hope for controlling the infection rather than eradicating the infection. The term 'eradication' is not used to describe the treatment of an individual patient with antimicrobial drugs, but should be reserved when discussing the elimination of a disease from a geographical zone (or, famously, smallpox from the world!).

To stand any chance of controlling vector-borne infections, an understanding of the biology of the vector itself is essential. As microbiologists, one may slip into thinking that expertise in the virus or bacterium is sufficient, but the behaviour of the vector is critical in devising strategies of control. For example, the sixty species of Anopheles mosquito that can transmit malaria to humans differ in their biting times and preference for breeding sites. The key features include knowledge of their breeding sites, biting behaviour, resting sites and host preferences (animal or human).

Wholescale spraying with insecticides such as DDT has been effective in the past, but the problem of the mosquito developing resistance to the insecticide has been steadily increasing. At first sight DDT appears to be attractive: stable (it remains active on the walls of sprayed rooms for days) and selectively toxic; weight for weight the toxicity is greatest against insects than against higher animals. Unfortunately, DDT has a half life of greater then 100 years and accumulates to toxic levels in the food chain. The environmental side effects of DDT have led to the more controlled use of DDT and other less potent chemical insecticides exemplified best in the development of impregnated bed nets. Of the various biological controls tested the use of a larvacidal

Female mosquitoes are blood sucking because of the nutritional demands of producing eggs. Males can live off nectar.

Prompt diagnosis and treatment of individual cases of vector-borne diseases will also serve to reduce the opportunities for transmission from infected hosts.

toxin has had the most success. *Bacillus thuringiensis* produces a pore-forming toxin that kills insect larvae and, being a member of the genus Bacillus, produces endospores which makes it feasible to spray large areas with a suspension of the organism in a suitably robust form.

The control of vectors is a national effort with the assistance of the World Health Organisation to co-ordinate such programmes. The financial resources and political commitment need to be maintained if any prolonged effects are to be realised. The return of mosquitoes to areas that have been previously cleared has been witnessed in many areas of the globe. Most vector control programmes aim to reduce the population size of the vector. Eradication being that much more difficult to sustain, it will usually stand better chances of success on islands rather than mainland areas. Table 11.2 compares control and eradication strategies.

### ■ 11.3.4 IMMUNO- AND CHEMOPROPHYLAXIS

**Immunoprophylaxis** is the administration of immune serum to people who are considered at acute risk of developing diseases against which there is a protective antiserum, for example, tetanus, botulism and diphtheria. Such protection is effective but of limited scope in that it protects the few people at risk for a limited period of time but has no effect on the control of the disease in the community. There is little, if any, distinction between the terms 'immunoprophylaxis' and 'passive vaccination'.

**Chemoprophylaxis** is the use of antibiotics in two circumstances:

1. to protect contacts of patients with infectious diseases from developing clinical disease, and
2. to provide temporary protection during the course of a medical intervention or surgical operation.

Again, chemoprophylaxis provides cover for the immediate contacts and cannot be employed to provide prolonged, widespread protection for communities. Antibiotics have not eliminated infectious disease or infections from the world, nor in fact have they significantly reduced their incidence. It may be argued that the prophylactic use of antibiotics has prevented infections in specific circumstances such as outbreaks of meningococcal meningitis in schools. In these circumstances close contacts (e.g. school children in the same school) are given antibiotics to prevent the disease developing. Immunisation will be inappropriate (assuming there is a suitable vaccine) because the time needed to develop immunity (2–4 weeks) will leave the contacts unprotected in the short term. Meningococcal meningitis can occur in clusters of people in a pattern

**Table 11.2 Comparison of control and eradication programmes for arthropod-borne diseases**

|  | Control | Eradication |
| --- | --- | --- |
| Objective | Reduce the incidence of the disease | Stop transmission and eliminate human infection |
| Duration | Indefinite | Limited period of time |
| Area of operation | Focused to key areas (urban environments, intense transmission) | All areas where transmission occurs |
| Long-term risks | Indefinite, recurring costs | Imported cases |

consistent with a common source outbreak. As vaccines are not available for all the serogroups of *Neisseria meningitidis* (antibodies against the polysaccharide capsules are protective) antibiotics are given to prevent the disease developing.

Another area where antibiotics have prevented infection is the prophylactic cover for surgical operations. For protection during surgery, the antibiotics need to be broad spectrum and given so that the peak levels are obtained during the operation. Whilst this has not abolished post-operative infections, the incidence of post-operative infections would undoubtedly be higher if such prophylaxis was omitted.

Table 11.3 summarises some of the factors that are likely to have an impact in reducing the infectious disease burden on a country. As stated earlier, the epidemics that scourged England were falling in both number and mortality before the advent of vaccines and antibiotics.

## ■ 11.4 ANTIBIOTIC RESISTANCE (BACTERIA)

The most pressing problem facing medicine is the growing number of micro-organisms that are now resistant to a wide range of antibiotics. The term **antibiotic resistance** describes the condition where bacteria have developed or acquired means to overcome the inhibitory effect of one or more antibiotics. A chilling example is multidrug-resistant *Mycobacterium tuberculosis*. Without the development of new antibiotics, thoracotomy (removal of the affected lobes of the lung) appears to be the only treatment option.

The mechanisms by which bacteria develops resistance to antibiotics are either due to **mutation** or **acquisition of resistance genes** from other organisms.

## ■ 11.4.1 MUTATION

Clinically, mutation leading to resistance has not been a major problem compared with the acquisition of resistance genes. In general, the effect of mutation will be to modify the target protein such that the binding affinity of the antibiotic is reduced. The protein will tolerate a certain loss of efficiency due to mutations, but constraints will limit the number and frequency of viable mutations in the active site. The affinity of binding will fall, but not such that the protein function is lost completely. If the binding site for the antibiotic is distinct from the active site of the protein, then there will be more scope for mutation to occur without significant loss of function. For example, the binding of an antibiotic on a ribosome might sterically hinder the correct binding of the tRNA and lead to misreading. If the antibiotic competed with the tRNA for the same binding site, then mutation of the tRNA binding site will limit the extent of mutation.

Mutational events reduce the potency of the antibiotic for the target site and this is observed by small increases in the concentration of antibiotic needed to inhibit the

**Table 11.3 Likely causes for control of certain infectious diseases**

| Reduced host susceptibility | Reduced transmisssion | Improved diagnosis | Improved treatment |
|---|---|---|---|
| Improved nutrition Improved housing | immunisation improved public health: water, food, sanitation, antibiotics | antibiotics | rapid diagnostic tests |

growth of the organism (minimum inhibitory concentration, MIC); for example, the MIC of *Staph. aureus* to penicillin may increase from 0.01 µg/ml to 0.05 µg/mL. By comparison, acquisition of resistance genes usually results in large increases in the MIC (e.g. from 1 µg/ml to >64 µg/mL).

An example of mutational resistance of clinical importance is the resistance of *Mycobacterium tuberculosis* to the aminoglycoside **streptomycin**. During the treatment of patients with tuberculosis, the proportion of bacteria that are resistant to the effects of streptomycin increases steadily with time due to mutations in the ribosomal binding site for streptomycin. The frequency of mutation of *Mycobacterium tuberculosis* is approximately 1 in $10^7$, meaning that one in every 10 million cells undergoes a mutational event that affects streptomycin activity. The mutation rate is fixed and therefore occurs irrespective of the presence of streptomycin. Thus, in patients taking streptomycin, the selective pressure results in the resistant mutants continuing to grow (unlike the susceptible cells), displacing the falling numbers of streptomycin susceptible cells until eventually all the bacteria will be resistant to streptomycin. If the antibiotic is withdrawn then the selective pressure vanishes and the susceptible cells will regain dominance because of the cost of streptomycin resistance in terms of reduced efficiency of ribosomal function. The key feature here is the constant rate of mutation; all that changes is the selective pressure on the organism by exposing the organisms to streptomycin.

Mutational resistance that causes problems for the treatment of a bacterial infection is largely restricted to *Mycobacterium tuberculosis*. Why?

The organism has a long generation time (2–4 hours) compared with 'fast' growing organisms such as *Esch. coli* with a generation time of 20–30 minutes and *Escherichia coli* has multiple copies of the rRNA gene whereas *Mycobacterium tuberculosis* has only one. Any mutation in one of the appropriate genes in *Escherichia coli* will be recessive to the unaffected gene copies whereas mutations in the single copy of the gene in *Mycobacterium tuberculosis* will be expressed.

### ■ 11.4.2 ACQUISITION OF RESISTANCE GENES (HORIZONTAL GENE TRANSFER)

Resistance genes are thought to exist in populations of bacteria that will be exposed to antibiotics in their natural environment. Bacteria in the soil will encounter antibiotics produced by fungi and *Streptomyces spp*. The strong selection pressures exerted by antibiotic use in agriculture (animal feeds, crop spraying, etc.), hospitals and the home (in countries where antibiotics can be bought freely at the pharmacy without prescription) has resulted in bacterial pathogens acquiring resistance genes from other organisms. The horizontal gene exchange of transmissible plasmids containing resistance genes means that effective antibiotic resistance is acquired in one step rather than waiting for mutational events to develop. The resistance genes have become widespread throughout different pathogens across the globe. Their mobility is conferred by becoming incorporated in transposons that, in turn, accumulate as gene cassettes in mobile plasmids called **multidrug resistance plasmids** (often abbreviated to **R plasmids**). They resemble the pattern observed with pathogenicity islands, clusters of genes all coding for related functions.

Mutational events may be the only option for organisms that do not readily exchange genes horizontally. Again, *Mycobacterium tuberculosis* is a good example. The cell wall structure of *Mycobacterium tuberculosis* has the extra mycolic acid/arabinogalactan layers that may prevent transformation, conjugation and transduction.

## ■ 11.4.3 RESISTANCE TO ANTIBIOTICS: CELLULAR MECHANISMS

### 11.4.3.1 Enzymatic inactivation of the antibiotic

A good example is the production of **β-lactamase**, an enzyme that hydrolyses the β-lactam nucleus of β-lactam antibiotics (penicillins). These enzymes may be found on the chromosome or on resistance plasmids and are expressed **constitutively** or as an **inducible** enzyme, only when β-lactams are present. The strategy adopted is exemplified in the Gram positive organism *Staphylococcus aureus* which produces inducible β-lactamase to be released extracellularly. In contrast the Gram negative *Escherichia coli* constitutively produces much smaller amounts of β-lactamase but sites it in the periplasmic space. In this way *Escherichia coli* chooses to only degrade β-lactam antibiotic that permeates through the porins in the outer membrane. Inactivating enzymes are also found that attack aminoglycosides and chloramphenicol.

### 11.4.3.2 Active efflux of antibiotic

The bacterium expends energy to actively extrude antibiotic from within the bacterial cell. The efflux pumps responsible may be able to pump more than one type of antibiotic (multidrug resistance efflux pumps).

### 11.4.3.3 Reduced uptake

An alternative to removing antibiotic once it has reached the cytosol is to mutate the mechanisms that are responsible for the uptake of the antibiotic in the first place. Such a strategy relies on active, i.e. energy expending, mechanisms of uptake. Passive diffusion will not apply.

### 11.4.3.4 Modification of drug target

Mutations will result in alterations of the binding sites in antibiotic target proteins. Penicillin binding proteins, for example, can reduce their binding affinity for β-lactam antibiotics. Alterations in the amino acid sequences of the ribosomal proteins can also result in reduced binding of the antibiotics that act on protein synthesis.

### 11.4.3.5 Overproduction of target

With antibiotics such as the sulphonamides and trimethoprim, competitive inhibitors of enzymes involved in the biosynthesis of folic acid, it is logical to see how bacteria have developed resistance to these drugs by simply overproducing enzymes. In the case of trimethoprim, the organism produces excess dihydrofolate reductase in order to overcome the competition from the drug.

Some of these mechanisms are produced by mutation, others are simply acquired through horizontal gene transfer. The origins of the resistance mechanisms lie in other organisms, presumably those that produce antibiotics themselves. When an organism is under prolonged pressure through exposure to antibiotics, in time, the appropriate resistance mechanism will eventually make itself available.

It is important to realise that bacteria can be 'resistant' to the inhibitory effects of antibiotics through mechanisms that are neither mutational events nor resulting from acquisition of exogenous resistance genes.

- Antibiotics have a particular spectrum of activity. Those organisms that are unaffected by a particular antibiotic may lack the target site or the mechanisms by which the antibiotic enters the bacterium.

MICROBIAL INFECTIONS OF HUMANS

- Even though several antibiotics are bactericidal (e.g. β-lactams), they can only exert this effect on actively growing bacteria; the organisms in stationary phase are not killed. This is a general phenomenon and applies not just to antibiotics but to disinfectants and other antibacterial conditions (starvation, drought). Endospores represent the extreme example of this strategy: completely inert structures that dominate in unfavourable conditions that are effectively unaffected by antibiotics.
- Dense collections of bacteria as biofilms offer protection partially through physical restriction and binding of the antibiotic so as to reduce the active concentration that can penetrate the biofilm.

## ■ 11.5 ANTIVIRAL DRUG RESISTANCE

When an antibacterial or antiviral agent acts at a single target site then resistance can develop through mutations at that site. The high mutation rates in RNA viruses will mean that such problems will occur more readily than with DNA-based bacteria and this is witnessed in HIV therapies. Again, as with antibiotic policies, certain manoeuvres can be employed to minimise resistance developing. These include the use of combinations of antiviral compounds, switching between two unrelated agents (although sequential use of antivirals has been criticised for simply promoting resistance rather than preventing it) and the use of high concentrations to prevent *any* viral replication ('knock out'). This, theoretically, prevents any viable mutants being released.

Resistance to antiviral agents will result from mutations in the viral genome. The modified proteins may act on cell events that occur upstream of the active compound (e.g. mutations in viral thymidine kinase will reduce its affinity of binding to the nucleoside) or mutations may occur in the target protein itself (e.g. retroviral reverse transcriptase). Whilst mutation is a frequent event in RNA viruses, and resistance develops frequently during the course of an infection, the spontaneous mutation events in a DNA virus like *Herpes simplex* will be of much lower frequency. Note that latent *Herpes simplex* virus does not express TK or DNA polymerase, hence the virus remains unaffected by the drug.

The laboratory testing for resistance to antiviral agents has a number of problems. Direct testing of the inhibitory concentrations of antiviral agent (**phenotypic testing**) gives a concentration that is active, at least under laboratory conditions. Phenotypic testing requires that the organism has been cultured and this is not possible for those viruses that have not yet been successfully cultured *in vitro*. Molecular mechanisms avoid this problem. Genotype testing can, in theory, be carried out on clinical samples as well as cultured virus. Polymerase chain detection for resistant gene sequences are increasingly available and are less technically demanding than phenotypic tests. The interpretation of both types of testing is somewhat arbitrary and, like antibiotic susceptibility testing in bacteria, will only indicate possible outcomes or provide frames of reference.

## ■ 11.6 SURVEILLANCE

In the latter half of the 1980s, the incidence of tuberculosis had increased to 'epidemic' proportions in a developed city, New York, in the most prosperous country in the world. How could such a well-recognised infectious disease against which an effective vaccine and a range of antibiotics are available spread unnoticed in such a city? The reasons illustrate how good public health infrastructure can be undermined if unsupported. Famously associated with poverty, tuberculosis had steadily increased in a place where HIV was endemic amongst people living in overcrowded, poor housing. The treatment of tuberculosis uses three antibiotics over 6 months and patients with pulmonary disease who are the infectious sources quickly become non-infectious following antibiotic treatment. If patients fail to complete the treatment, they will relapse into infectious cases and antibiotic resistance may result. Of the measures taken, DOT (**direct observation of therapy**) was crucial to the management of the epidemic. DOT targets the source by effecting a reduction of the proportion of infectious cases. One of the reasons for the return of tuberculosis was the reduced funding to public health departments responsible for monitoring (surveillance) and treating the disease. Partly this was a consequence of years of success in reducing the incidence of tuberculosis such that complacency had set in. Not surprisingly the cost of bringing the epidemic under control was more than the savings made in the first place.

## ■ 11.7 FUTURE EPIDEMICS

Epidemics have been a regular feature of human existence throughout history, so it is easy to state with confidence that they will continue to be so. The reasons for such a gloomy outlook are not difficult to find (see Box 11.3). As populations have increased in size, particularly in cities, so has the incidence of infectious disease. The proportion of people infected by certain infectious diseases (tuberculosis is again an example) may fall, but the absolute numbers of people infected will still increase. The present population forecasts indicate increasing numbers of large (greater than 10 million people) cities, mostly in ports and concentrated in developing countries. Other than providing a critical population threshold to sustain diseases like measles, sexually transmitted diseases and the common cold, large cities need global commerce to support them, as well as attracting migration, which both increase the inward and outward flux of people, goods and microbes. The public health infrastructure needs to be appropriately sized and funded to prevent the resurgence of old infections that, although once under control, will readily return without continued control. The expansion of the areas of known infectious disease is a result of expanding urban slums into the country such that the associated risks of deforestation and monoculture of crops all result in disturbing zoonotic cycles. It is also depressing to note the regular appearance of epidemics that accompany war, a habit that, far from diminishing, seems to be on the increase and risks employing biological weapons. Although poverty is famously linked with poor health, notably tuberculosis, wealthy nations are not excluded from risk as advances in medicine keep more people alive with increasing opportunities for infections.

International surveillance will become increasingly important as the world shrinks through globalisation in order that infectious diseases can be tracked. The bioterrorist threat, exemplified by the crippling cost of sending a few samples of anthrax spores

The value of quarantine measures to screen imported animals for infectious diseases needs consideration. The ancient practice of retaining ships in port for 40 days ('quarantina') applied to all trade, not just to animals, and was driven mostly by a need to prevent the spread of plague. With the spatial isolation of an island, quarantine can prevent the introduction of a disease. With increasing international travel by air, the principle has been undermined. The building of a tunnel under the English channel to connect the United Kingdom with the rest of Europe means the isolation is breached again.

---

### ■ BOX 11.3 EXAMPLES OF CHANGING PATTERNS IN INFECTIOUS DISEASE

- Previously unrecognised diseases: Lyme disease (*Borrelia burgdoferi*), *Helicobacter pylori* gastritis.
- Increasing zone of disease: increasing malaria and Dengue due to the spread of mosquito vectors into new areas.
- Resurgence of 'old' infections: epidemic wave of diphtheria in Europe affecting the old communist bloc.
- Apparent synergy in tuberculosis with HIV in AIDS patients.
- Increasing incidence of drug-resistant tuberculosis.
- Food-borne infections increasing with globalisation of food products.

through the post, is a further call to adequate microbiological infrastructure and expertise. The factors that are important in protecting the public from biological weapons are similar to those that are considered in all microbiology laboratories. The comparisons are given in Box 11.4.

The need for microbiologists is obvious but should that idea need reinforcement then it is worth noting that the cost of eradication of poliomyelitis in the USA has paid for itself every 26 days since 1977. The financial losses to a country incurred through infectious disease every year may yet make governments take notice.

Having described the human activities that promote infectious disease, what are the features of the organisms that will continue to create problems? RNA viruses are the organisms with the highest mutation rates. It can be expected that such organisms will continue to cause epidemics, just as HIV has been the most frightening of recent times because of the genetic variation that occurs through inaccurate transcription and genetic reassortment. HIV has demonstrated the potential impact that an RNA virus can have if it is able to cross the species barrier (from apes to humans). The risk of mutational events resulting in greater virulence in bacteria and fungi is less than for RNA viruses, but the increasing recognition that toxins and pathogenicity islands are coded for on mobile genetic elements means that bacteria will continue to present threats. The increase in antibiotic resistance genes through horizontal gene transfer over the last 50 years reminds us that we have not yet overcome the existing microbial threats.

---

### ■ BOX 11.4 HAZARD GROUPS

In order to protect the staff in microbiology laboratories as well as the public and the environment, micro-organisms have been categorised into four Hazard Groups. The groups, in increasing order of hazard, are as follows:

Group 1: Unlikely to cause human disease.

Group 2: Can cause disease and may be a hazard to employees.
Unlikely to spread to the community.
Effective prophylaxis or treatment available.

Group 3: Can cause human disease.
Risk of spread to the community.
Effective prophylaxis or treatment available.

Group 4: Can cause severe human disease.
Likely risk of spread into the community.
No effective prophylaxis and treatment.

The micro-organisms considered the most effective biological weapons are chosen because they:

- are easily disseminated into the public or are highly infectious (readily transmitted from person to person).
- cause high mortality.
- require special action for public health control.

The organisms mostly fall into Hazard Group 4. Viruses such as *Variola major* virus (smallpox) and Ebola and Marburg viruses are not easily treated with antiviral agents but there are effective vaccines available (although not routinely used). *Bacillus anthracis* and *Yersinia pestis* (the aetiological agent of plague) can be treated with antibiotics.

## ■ 11.7.1 RABBIT OR THE TORTOISE?

Headline infections such as AIDS and cholera, in which epidemics sweep through populations with dramatic consequences, will eclipse the fact that the most frequent cause of death due to an infectious micro-organism is tuberculosis. With chronic but steady progression, the disease is quietly spreading through developing countries such that almost 2 billion people are thought to be infected, 2 million of whom die every year. *Mycobacterium tuberculosis* is a strict human pathogen and has no other reservoir. It is interesting to compare how *M. tuberculosis* has obtained this position of dominance with organisms like *Vibrio cholerae* or HIV. The organism has a cell wall structure that is distinct from the archetypal Gram positive and Gram negative bacteria; instead it is dominated by a thick layer of complex glycolipids. The organism has one of the slowest growth rates of all pathogens. It has no known protein exotoxin or type III secretion systems or recognised pathogenicity islands, and horizontal gene transfer appears to be rare. The pathology instead results mostly from the host responses. Whereas most bacterial pathogens are driven from carbohydrate metabolism, *M. tuberculosis* utilises lipids as principle metabolic fuels and then uses lipid products to build the cell wall. Lipid-based compounds are manufactured to act as immunomodulatory molecules that confer the ability to persist for life in the host, residing within macrophages. Spread via aerosols from infectious cases, the organism adopts a common mode of transmission. What distinguishes *M. tuberculosis* from another aerosol-transmitted infectious disease like measles is the duration of infection (lifelong). By adopting a 30 year latency period within the host, the infectious cases are virtually guaranteed to encounter a susceptible host at some stage. Although the difficulties in genetically manipulating mycobacteria and having no suitable animal models of latency have delayed research on putative virulence factors, an understanding of disease epidemiology suggests that the long-term approach to parasitism is an important feature. Whatever the mechanisms, the genus Mycobacterium appears to be well suited to the parasitic existence both in humans and other animals; *Mycobacterium bovis* is the major pathogen of animals.

## ■ SUMMARY

Epidemiological analyses have become increasingly important in the control of infectious diseases. The concept of the reproductive rate has proven to be a useful starting point for understanding the epidemiology of infectious diseases and permits estimates of the percentage vaccination uptake that is required to eliminate an infectious disease. Similarly, the concept of the persistence threshold number can predict the persistence or periodic epidemic waves of an infectious agent.

An increasing number of vaccines, both in range of targets and vaccine type, have been developed and still they provide the most effective means of controlling infectious diseases in populations. The use of antibiotics is becoming increasingly threatened through the development of resistance mechanisms by numerous pathogenic micro-organisms.

The patterns of human behaviours are powerful influences in promoting opportunities for new infectious disease. As human behaviour continues to change local and global ecology, new infections and epidemics will continue to appear. Thus the role of public health measures remain as important today as they were in the early nineteenth century, but international surveillance is likely to be the next crucial strategy in attempting to minimise the developing microbial threats to human health.

## RECOMMENDED READING

Ciba Foundation Symposium, Number 207 (1997) *Antibiotic Resistance: Origins, Evolution, Selection and Spread*, Wiley, Chichester, UK.

Davies, J. (1994) Inactivation of antibiotics and the dissemination of resistance genes. *Science* 264, 375–82.

Nikaido, H. (1994) Prevention of drug access to bacterial targets: permeability barriers and drug efflux. *Science* 264, 382–8.

Noah, N. and O'Mahony, M. (eds) (1998) *Communicable Disease Epidemiology and Control*, John Wiley & Sons Ltd, Chichester, UK.

Nokes, D.J. and Anderson, R.M. (1988) The use of mathematical models in the epidemiological study of infectious diseases and in the design of mass immunisation programmes. *Epidemiol. Infect.* 101, 1–20.

Spratt, B.G. (1994) Resistance to antibiotics mediated by target alterations. *Science* 264, 388–93.

## REVIEW QUESTIONS

*Question 11.1*   What are the mechanisms by which bacteria resist the action of antibiotics?

*Question 11.2*   Define the reproductive rate ($R_o$). How does it differ from the net reproductive rate ($R$)?

*Question 11.3*   What are Arbovirus infections and why are they difficult to control?

*Question 11.4*   What is passive vaccination?

*Question 11.5*   Compare and contrast control and eradication programmes.

*Question 11.6*   List factors that may be thought responsible for the development of 'new' epidemics or infectious diseases.

# INDEX